PRACTICAL METEOROLOGY

PRACTICAL METEOROLOGY

DREW JOHN, FREDERIC DREW

GYAN

Published by

Gyan Publishing House
5, Ansari Road
Daryaganj, New Delhi-110002
Phone: 011-47034999, 9811692060
E-mail: books@gyanbooks.com

Distribution Network
gyanbooks.com
India, USA, Canada, UK, Australia

ISBN: 978-93-7169-476-6 (PB)

First Published, 1860

2nd Impression 2025

Printed at: Gyan Press, Delhi.

PRACTICAL METEOROLOGY
Author: DREW JOHN, FREDERIC DREW

Pl. I.

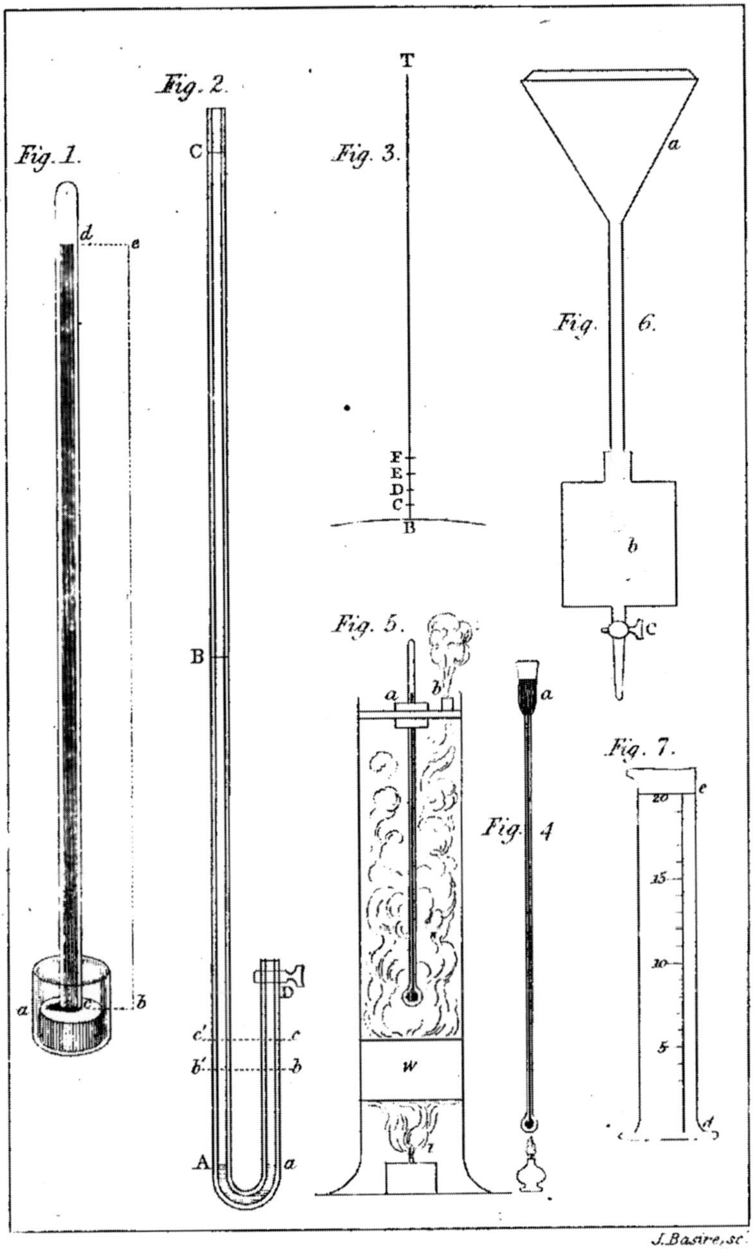

J. Basire, sc.

London: John Van Voorst, 1855.

PRACTICAL

METEOROLOGY.

BY

JOHN DREW, PH.D., F.R.A.S.

SECOND EDITION,

EDITED BY

FREDERIC DREW.

LONDON:

JOHN VAN VOORST, PATERNOSTER ROW.

MDCCCLX.

PREFACE.

In the year 1847, the author, having resolved to undertake a series of observations on the climate of Southampton, made inquiries for some treatise on the subject of Meteorology which should assist him in the choice and use of instruments; such a work was not to be found, and the required information was obtained from intercourse with some of the most eminent meteorologists at home and abroad, together with an attentive study of the Greenwich Observations, and of the records of various scientific Societies.

Three years later the author was invited to

assist at the formation of the British Meteorological Society; and being anxious, as a Member of the Council, to promote its object, he published in a scientific periodical, a series of papers "On the Instruments used in Meteorology, and on the Deductions from the Observations," which were extensively circulated by the Secretary among the Members, of whom several, whose position added weight to their representations, urged the author to republish the series in a more permanent form; the result of seven years' experience as an observer and student is the present volume, which professes to teach "how to observe" meteorological phenomena.

The author apprehends that every possessor of a Barometer, Thermometer, or Hygrometer, will here find ample directions for its advantageous use; while the general reader will learn from this work the progress which has been made and the appliances which have been brought to bear in the pursuit of the laws which regulate atmospheric changes.

The work consists of three Parts :—

Part I. Introductory: on the Laws of Heat as affecting Atmospheric Changes.

Part II. Instruments of Observation described. Deductions from Observations on the Thermometric, Hygrometric, Barometric, and Electric Condition of the Air.

Part III. The present state of Practical Meteorology in this Country. Description of the Photographic Registration of Phenomena at the Royal Observatory, Greenwich.

The various Tables which are scattered throughout the work will be found to contain every requisite for the reduction of observations. For facility of reference the principal of these are here enumerated :—

Comparison of Thermometric Scales, p. 285.

Diurnal Range of Temperature, p. 63.

Diurnal Range of Evaporation, p. 168.

Tension of Aqueous Vapour, p. 292.

Degree of Humidity from Readings of the Dry- and Wet-bulb Thermometers, p. 296.

For measuring Heights with the Barometer, p. 215.

Weight of Vapour in a Cubic Foot of Air, p. 293.

Weight of a Cubic Foot of Air, p. 295.

Correction of the Barometric Reading for Temperature, p. 297.

Surbiton, Kingston-on-Thames,
March, 1855.

PREFACE TO THE SECOND EDITION.

In this edition of "Practical Meteorology," I have endeavoured to give an account of every useful instrument that has been invented since the first was published, and, by a slight change in the arrangement of the sections of the book, to put more clearly the information it is intended to convey; I have, in fact, tried to make it bear the same relation to Meteorological Science as it is now, that, from the way in which the first edition was received, it may well be supposed the book had five years ago.

F. D.

Museum of Practical Geology,
Feb. 2, 1860.

CONTENTS.

PART I.—INTRODUCTION.

Part II.—OBSERVATION AND DEDUCTIONS.

APPENDIX.

PRACTICAL METEOROLOGY.

PART I.

INTRODUCTION.

1. **Definition.**—A knowledge of the causes which produce, and of the laws which regulate atmospheric changes, is the object of the science of Meteorology.

2. **General view of the relation of the atmosphere to the earth.**—To realize the extent and the proportion which the atmospheric envelope bears to the earth, let us imagine her to be viewed from a neighbouring globe. Before the spectator would float a massive sphere, nearly 8000 miles in diameter; the whole of its illumined portion would never be free from clouds and vapours, which would draw an impenetrable veil over certain parts of the continents and oceans, so that the outline of these formations would only be seen occasionally; in some parts, however, more distinctly than in others. The presence of an atmosphere would be indicated when the earth

occulted a star, that is, passed over it as we observe the moon does in traversing her orbit: the light of the star would fade by degrees before the body of the earth reached it, the diminution commencing when the star entered the atmosphere; but whether it would be affected when it first approached its limits is questionable. The utmost extent to which the air reaches is $\frac{1}{200}$ of the earth's diameter, but probably the effect would be imperceptible till the star had advanced towards the earth's disc within $\frac{1}{2000}$ of her diameter. The observer, if he had the means of ascertaining, would see that as the earth is not a sphere, but a spheroid, the equatorial diameter exceeding the polar by $26\frac{1}{2}$ miles, so the atmosphere would be in form a spheroid, but not exactly similar to that of the earth which it surrounds; since, as the force of gravity is at its maximum at the poles, there the extent of the atmosphere would be less, or it would be more compressed or more dense, seeing that air, like all other matter, would be under the influence of gravitating attraction.

The shadows of the mountains would not be absolutely without light like those on the moon, but would be partially enlightened by the refractive and reflective powers of the atmosphere; the boundary of light and shade between the bright and dark portions of the earth's surface would show a blending of one into the other; in other words, day and night would be separated by an extensive twilight, arising from the dispersion of the sun's rays by the upper strata of the atmosphere, on which his light continues to fall when he has set to the surface immediately below them. It is indeed from the reflective and dispersive properties of the atmosphere as regards light, that the shadows of objects on the earth are not a deep black; every particle has this power of scattering the rays of the sun and affording a secondary illumination.

Such would be the appearance of that "aërial ocean," on the bottom or shoals of which man and all other animated beings find the means of their subsistence: general views would be attained from such a point of observation as that we have supposed; for more particular investigation we must return to our position on the surface of the earth itself.

3. **Composition of the atmosphere.**—The atmosphere is found, when analysed, to consist for the most part of two gases, oxygen and nitrogen; not chemically combined, but mechanically united. Oxygen supports combustion energetically, and in it the vital functions are carried on, but far too rapidly to consist with the well-being of living

creatures. Nitrogen extinguishes flame and destroys animal life. If a piece of phosphorus be burnt in a closed jar of air over water, it will combine with the oxygen and form phosphoric acid, which, being taken up by the water, will leave the nitrogen in the glass; into this insert a lighted taper and it will be extinguished—a small animal, and it will expire.

In addition to these two gases, which form the bulk of the atmosphere, a small portion of carbonic acid gas is always present; it may be detected by exposing to the air a saucer of lime-water, which will soon become turbid from the formation of insoluble carbonate of lime.

The air, at all times, has diffused through it a certain amount of water in the form of vapour, the quantity of which is variable.

4. **Proportion of the constituents.**—From the recent very careful analyses of air by M. Dumas, the following proportions have resulted:—

	Air by weight.	Air by volume.
Oxygen	23·10	20·90
Nitrogen............	76·90	79·10
	100·00	100·00

The proportion of the constituents of air freed from moisture, including those of which traces only are indicated, are as follows in 10,000 volumes:—

Composition of dry air by volume.

Nitrogen	7,912
Oxygen	2,080
Carbonic acid	4
Carburetted hydrogen .	4
Ammonia	trace.
	10,000

5. **Air at high elevations.**—It is found that in air, whether taken from near the level of the sea or at the greatest elevations above it, to which, in the pursuit of science, we have attained, the proportions of oxygen and nitrogen are the same. The latest experiments by which this conclusion has been confirmed, were performed by Dr. Miller, of King's College, London, on the air brought down from various heights in one of the balloon ascents, projected by the Council of the British Association for the purpose of scientific investigation.

Dr. Miller thus describes the method and the results of his analysis, on samples of air brought down on August 26, 1851 *.

The recipients for the air were wide glass tubes about 5 cubic inches in capacity; they were filled with the specimens to be examined by simply opening and then closing a stopcock, the altitude being determined by an observation of the

* Philosophical Transactions, 1853.

barometer at the moment; the tubes were, within twenty hours after the air had been collected, hermetically sealed, and the proportions of oxygen and nitrogen determined with great care by detonation with hydrogen in "Regnault's Eudiometer." The volumes of oxygen found in the air collected at different altitudes are given in the following Table :—

	Altitude.	Vol. of oxygen.
Air collected at King's College	....	20·920
Tube 2	13,460 ft. ...	20·888
Tube 3	18,000 ft. ...	20·747
Tube 1	18,630 ft. ...	20·888

Hence it is found that the composition of the atmosphere, as Gay-Lussac had announced as the result of his experiments at a time when the methods of gaseous analysis were less perfect than at present, exhibits no sensible difference, whether taken from near the surface, or at the greatest heights accessible to man.

The presence of carbonic acid was distinctly shown by a film of carbonate of lead upon a solution of the subacetate, which was introduced to a portion of the air confined over mercury.

6. **Ozone.**—Under certain electric conditions of the atmosphere, a principle would seem to be developed, to which the discoverer, Dr. Schönbein

of Bâle, has given the name of "*ozone*;" to this reference will be made hereafter.

7. **How the elements are combined.**—That the air is not a compound of oxygen and nitrogen, but only a mixture of these two gases, has been argued by a laborious investigator (M. Regnault) on the following grounds.

α. By the law of chemical composition, gases combining chemically in volumes are found to be in certain definite proportions; the following is the nearest approach to the proportionate quantities of each gas in atmospheric air:—

$\frac{1}{5}$ of oxygen—or oxygen...........	20·00
$\frac{4}{5}$ of nitrogen—or nitrogen.........	80·00
	100·00

These numbers deviate more from the results of direct analysis than can be due to errors in the experiments, which are all consistent in giving 20·9 and 79·1 as the per-centage of oxygen and nitrogen.

β. When gases combine, heat is always given out; but there is no appreciable change of temperature when oxygen and nitrogen gas are mixed experimentally in the proportions constituting common air, although the union produces a gas which is identical with the atmosphere.

γ. Water which has remained for a long time

in contact with air, always holds a certain amount of it in solution. If the air were a chemical compound, that derived by very nice analysis from water ought to exhibit precisely the same properties as atmospheric air, and the proportions of the gases ought to be the same. If, on the contrary, the gases are only mechanically combined, seeing that water will dissolve ·046 of a volume of oxygen, and only ·025 of a volume of nitrogen, the one gas will be in excess in air derived from water. The careful analysis of such air has clearly indicated this fact. Hence we conclude that oxygen and nitrogen are intermingled, and not combined, in the compound gas which constitutes the air we breathe.

8. **Mechanical properties of air.**—Of the mechanical properties of air we shall mention weight and elasticity.

Weight.—Its weight becomes evident when a receiver, over the top of which has been strained a piece of bladder, is exhausted by means of an air-pump: the bladder will be found to curve downward as the exhaustion proceeds; and, as the air underneath is pumped out, it will at last burst with a loud report.

M. Regnault has lately most carefully determined the weight of air deprived of carbonic acid and aqueous vapour. With a pressure equal to that of 30 inches of mercury, and at a temperature of 60°, 100 cubic inches of air have been found to weigh 30·82926 grs.

The mean pressure of a vertical column of air at the sea-level is usually estimated as equal to 30 inches of mercury, or as 15 lbs. to every square inch of surface. If, in the experiment mentioned above, the mouth of the receiver was 6 square inches in area, on the supposition of a perfect vacuum being formed underneath, the pressure on it would be equal to $15 \times 6 = 90$ lbs.

Elasticity.—The property of elasticity—by which contraction of bulk occurs on applying pressure, and expansion on its removal—is well illustrated by the air-gun. A large quantity of air is forced into a small space; on its being allowed to escape, it urges a bullet before it with considerable force, in attempting to resume its original volume.

9. **Aqueous vapour exists in combination with the atmosphere.**—The vital air—the lumen spirabile cœli—does not consist only of the admixture of oxygen and nitrogen which has been described; co-existing with it is another gaseous fluid, which is in fact the vapour of water. We may with truth consider the globe surrounded with two atmospheres, the one of air, the other of aqueous vapour, not chemically combined, but commingled

or mechanically united; each being, as it were, diffused through the pores of the other: at all times and in all places this union exists, though the amount of the aqueous vapour is variable to an extreme degree, depending indeed on the temperature of the air, and on the neighbourhood of large quantities of water. The exact quantity which exists in the air at any particular instant may be determined most accurately. The changes which we observe in climate or in weather, may in general be traced to the preponderance or deficiency of the vapour of water in the air, and to the interference of the laws to which it is subject, with those which would obtain in an atmosphere of perfect dryness.

10. **Dalton's theory of the composition of the air.**—The theory of Dalton with respect to the constitution of the compound atmosphere was, that, of the various elastic fluids which compose it, the particles of one have neither an attractive nor a repulsive power towards those of another, and that the weight or pressure of any one particle of any gaseous mixture of this sort arises solely from the particles of its own kind. According to this view, it is possible that oxygen, nitrogen, and aqueous vapour may exist together under any pressure, while each of them is diffused throughout the whole space allotted for all. Whether or

not this is absolutely correct, we shall show hereafter that it is possible to appropriate to the air, and to the water in it, the portion of the entire pressure of the two united gaseous fluids which is due to each.

On the supposition of Dalton, Dr. Henry calculated the pressure due to each constituent of the atmosphere, in a mean state as regards moisture, to be in such proportion as the following:—

	inches of mercury.
The nitrogen gas exerts a pressure equivalent to	23·36
The oxygen gas	6·18
The aqueous vapour	0·44
The carbonic acid gas	0·02
	30·00

11. Barometer explained.—The instrument by which we arrive at a knowledge of the amount of atmospheric pressure is the Barometer, the mode of action of which must be explained at this stage of our progress. The details of construction will be omitted till we enter upon the consideration of the barometric condition of the air more fully in Part II.

Fill a glass tube 33 or 34 inches in length, and closed at one end, with mercury, invert it in a vessel of the same liquid, and a barometer will be constructed. The column of mercury *d c* (Plate I. fig. 1) will descend until its weight exactly balances the pressure of the atmosphere on the surface *a c* which is open to its influence. A scale measuring the height of the top of the column above the surface of the liquid in the vessel, which height will be the same whatever the diameter of the tube, will show, from time to time, the variation in the pressure of the air vertical to the place of observation. The mercurial column will descend as the pressure or weight of the air decreases, and will rise as the pressure becomes greater, or as the air becomes more dense.

12. Ratio of decrease of atmospheric pressure.—It will be readily granted that air near the level of the sea, having to bear the weight of the entire mass above it, is denser than any superior stratum. The illustration of Pascal who compared the atmosphere to a mass of wool, the lower parts of which having to bear the superincumbent mass will be compressed into a smaller space than the upper, will serve to render the fact clear to every comprehension. To demonstrate the ratio of decrease as we ascend above the sea-level, will require the aid of mathematical formulæ.

13. Mariotte's law.—Before entering on this investigation, we must fully apprehend the bearing of Mariotte's law, namely, that the density of a gas, other things being equal, is directly as the pressure that confines it, as is also its elastic force; while the volume occupied by a certain mass is *inversely* proportional to the pressure; and therefore the density and the elastic force of a given mass are inversely as the volume. Thus, if a portion of air which fills two cubic feet of space be compressed into the space of one cubic foot, its density will be doubled, and its elastic force made double of what it was originally; or, again, if we take a certain volume of air under a given pressure, that volume will become twice, three times, or ten times less, if the pressure become two, three, or ten times greater.

Mariotte demonstrated this law in 1650, and late experiments have shown that it holds good, as regards the air, when it has been expanded to 300 volumes, and also when compressed to $\frac{1}{25}$th of its primary value: to this extent have the experiments ranged.

14. Experimental proof.—That the density of the air is proportional to the force which compresses it may be thus shown:—Let a siphon tube be taken, C B A D, Plate I. fig. 2, of which the longer end, not less than 90 or 100 inches, is open, and the short end supplied with a stopcock D, which may be opened or closed at pleasure; by pouring in a small quantity of mercury, the communication between the two ends will be cut off. Let A *a* be the level of the mercury in the two tubes; close the stopcock D, and it is plain that the air contained in *a* D will be of the same density as the external air. Observe the height of the mercury in a barometer adjacent at the time of the experiment, and pour in at C as much mercury as, when measured from the level of *b*, the point to which it will have risen in the short arm, shall equal the height of the mercury in the barometer; we shall now have the column B *b'*, equal to the pressure of another atmosphere, tending to compress the air in the tube, which will be found to occupy the space *b* D, just half of the original space *a* D; that is to say, its density is doubled; in other words, the air at first was of such a density as resulted from its having to bear the pressure of the atmospheric column above it; now, the additional weight of a column of mercury equal to such pressure, has halved its volume and doubled its density. By adding mercury to the same amount as before, so that the height c' C $= 2b'$ B, the air will become subject to the pressure of three atmospheres, and will occupy *c* D, one-third of the original space, and therefore will be of three times the original density. Hence we find that in all cases the density is directly proportional to the compressing force, and the volume will vary inversely as that force. Now, to apply this law to the elucidation of the variation in the pressure of the air on ascending to different heights above the level of the sea,—

15. Ratio between height and pressure explained.—Let us suppose an ascent to have taken place from the level of the sea, at which the barometer indicated a pressure of 30 inches of mercury, the temperature shown by the thermometer being 60° throughout; the decreasing pressures at each mile of direct ascent would be as follows:—

Pressure in inches.	Height in miles.
30·000 = a	0
24·817 = b	1
20·530 = c	2
16·983 = d	3
14·049 = e	4

Here the heights are evidently a series of numbers in arithmetical progression, whose first term is 0, and common difference 1.

The pressures may be shown to be in geometrical progression, for $a:b::b:c::c:d::d:e$; or calling r the ratio between a and b, b and c, &c.,

$$\frac{a}{b}=\frac{b}{c}=\frac{c}{d}\ \&c. = r.$$

The first term in the above series is 30, and the ratio ·827: here $30 \times \cdot 827 = 24{\cdot}817$, and $24{\cdot}817 \times \cdot 827 = 20{\cdot}53$, and so on.

To exhibit the same law of decrease in pressure, or increase in volume, of the air in another point of view, let 1 represent a volume of air at the sea-level; then we shall have

Pressure in inches.	Height in miles.	Volume.
30	0	1
15	2·705	2
$7\frac{1}{2}$	5·41	4
$3\frac{3}{4}$	8·115	8
$1\frac{7}{8}$	10·82	16
$\frac{15}{16}$	13·525	32
$\frac{15}{32}$	16·23	64

Here the heights form an arithmetical series, of which the first term is 0, the common difference 2·705; the volumes a geometrical series, the first term 1, and the ratio 2; the pressures (inversely as the volumes) another geometrical series, the first term of which is 30, and the ratio $\frac{1}{2}$.

16. Proof of the same.—The following is perhaps the simplest proof that can be given of the law of the decrease of pressure:—

Let B T, Plate I. fig. 3, represent a cylindrical column of the atmosphere, pressing vertically on the earth's surface at B; let it be supposed to be divided into a number of layers so thin that the density may be considered uniform throughout the extent of each. Let BC, CD, DE, EF be four of these layers, equal in thickness. Now, the weights of these equal volumes, BC, CD, DE, EF, are directly as their densities, and the density of each is directly as the pressure upon it; therefore the weights, which let us call a, b, c, d, are directly proportional to the weight of air at C, D, E, F, respectively; that is, putting t for the weight of air above F, a, b, c, d are as

$$b+c+d+t, \quad c+d+t, \quad d+t, \text{ and } t,$$

$$\therefore \quad \frac{a}{b+c+d+t}=\frac{b}{c+d+t}=\frac{c}{d+t}=\frac{d}{t},$$

that is, there is a fixed ratio between the weight of a stratum of air, of a certain thickness, but at any level, and the weight of air incumbent on it.

Now if we call this ratio $\frac{1}{n}$, the pressures at the successive stations will be

at F $\quad t$;

at E $\quad t+\frac{t}{n}$, or $t\left(\frac{n+1}{n}\right)$;

at D $\quad t\left(\frac{n+1}{n}\right)+\frac{t}{n}\left(\frac{n+1}{n}\right)$, or $t\left(\frac{n+1}{n}\right)\left(\frac{n+1}{n}\right)$;

at C $\quad t\left(\frac{n+1}{n}\right)\left(\frac{n+1}{n}\right)+\frac{t}{n}\left(\frac{n+1}{n}\right)\left(\frac{n+1}{n}\right)$, or

$$t\left(\frac{n+1}{n}\right)\left(\frac{n+1}{n}\right)\left(\frac{n+1}{n}\right);$$

which quantities are a geometrical series with $\frac{n+1}{n}$ for the ratio. Therefore, when the heights ascended are, as in this case, in arithmetical progression, the pressures are in geometrical progression.—Q. E. D.

17. Limit of the atmosphere.—From the principle now established it may be considered that although the density of the atmosphere may be continuously decreasing, yet that it never amounts to 0, and therefore that the atmosphere is unlimited in extent, or extending through space in infinite divisibility. But we must be careful how we thus reason; laws regulating matter so attenuated may be yet unknown; it may be that the repulsive power of the particles of the atmosphere, which in all gases is great, is diminished, as the rarity increases, to such an extent that the weight or gravitation of the particles at last balances it, and prevents further divergence; and thus at last it may have an upper surface like a liquid. We are led to conclude, moreover, from the phenomena of refraction, that about 45 miles is the extent to which the atmosphere reaches above the surface of the earth.

The exact ratio between the heights ascended and the pressure will hold good only on the supposition that the temperature throughout remains the same, which is in fact never the case.

18. Heat a powerful agent in atmospheric changes.—Most of the changes which we observe in climate or in the weather arise immediately from heat, which, among its other effects, produces a constant variation in the amount of aqueous vapour in the air; whence arises the interference of the laws regulating a mixed atmosphere, with those to which dry air would be subject. We shall find, moreover, that various meteorological phenomena are brought about by heat absorbed or radiated by the earth; and as in the construction of instruments of observation the effect of heat on the materials of which they are made must be taken into account, we shall facilitate our progress by explaining the laws of heat, at this stage, for future reference.

19. Laws of heat: *Expansion.*—Experiments show that bodies, with few exceptions, increase in volume by the application of heat, the increase taking place regularly on the application of equal increments of heat, and recover their original size when reduced to the initial temperature: thus if the temperature of an iron bar be raised 5°, and thereby increased in length ·1 of an inch, by elevating the temperature 5° more, it will become ·2 of an inch longer than at first.

Liquids are amenable to the law of expansion: insert a tube of glass through a cork into a Florence-flask filled with water; apply heat, and the water will rise in the tube—the higher as it becomes warmer.

The thermometer, the instrument by which we

measure any degree of heat below that of a fire or a furnace, is constructed on the principle of mercury expanding, or rather dilating, in equal amounts for equal increments of heat.

20. Exception to the law.—The exception to the law of expansion by heat, and contraction by the loss of it, is water, which obeys it however till cooled down to 39°·5; if the process of cooling is continued below that point, it begins to expand, and, in a state of perfect stillness, it will continue to do so, without freezing, for 20° below the freezing-point: when frozen it is still lighter than when liquid, hence ice floats on the surface and is readily exposed to thaw on the return of warmth to the air: had water followed the usual law, the ice and coldest water would have accumulated, year after year, at the bottom of rivers and lakes, which the heat of the summer sun would never have reached.

21. Amount of expansion.

Table of the linear expansion of certain solids by heat.

Dimensions which a bar takes at 212°, whose length, at 32°, is 1·0000000:

Cast iron	1·0011111	the increase being 1 part in 900
Cast brass	1·0018750	1 part in 533
Zinc	1·002942	1 part in 340
Glass	1·0008613	1 part in 1248

Expansion in volume by heat—

of Mercury from 32° to 212° 0·018099 or 1 part in 55
Water from 39° to 212° 0·043320 or 1 part in 23

22. **Expansion of gases by heat.**—Gases expand by heat; and air, which is a gaseous compound as we have seen, does the same. The subject of the expansion of gases has been investigated by Dalton and Gay-Lussac, and more recently by Regnault, who has given $\frac{366\cdot5}{1000}$ or $\frac{11}{30}$ as the amount of the expansion of a volume of air in passing from the freezing to the boiling-point of water; that is to say, a quantity of air equal in volume to 1000 cubic inches at 32°, will expand to 1366·5 cubic inches at 212°, or by an increase of 180° of heat; hence, as it has been proved that the expansion is the same for each additional degree of heat, the expansion will be equal to 0·002036, or $\frac{1}{491}$ for each degree; in other words, 491 cubic feet of air at a temperature of 32° will become

492 at 33°
493 at 34 &c.
490 at 31
489 at 30
488 at 29 &c.

23. **Laws of heat:** *Conduction.*—Bodies lose or

acquire heat by conduction; the particles nearest the source acquire heat and transmit it to the nearest to them; these to the next, and so on till the whole mass has attained the heat of the surrounding medium. If the temperature of this medium, suppose the air, be reduced, the surface of the heated body radiates forth its heat, and, as it cools, borrows heat from the interior till equilibrium is restored.

According to the greater or less rapidity with which heat is diffused throughout a substance, so is it said to be a good or a bad conductor of heat; metals are good conductors of heat, gases, fluids and earthy matters scarcely conduct heat at all.

24. **Illustrative experiment.**—The different conducting powers of solid bodies may be shown experimentally thus: into a rectangular tin vessel insert, through apertures in the side, small cylindrical bars, equal in size, of various substances, which have been previously dipped in melted wax, from which they will have acquired a coating of that material of equal thickness throughout. Fill the vessel with boiling water, and the wax will melt soonest on the bar of that material which is the best conductor of heat. By making the bars of iron, lead, porcelain, &c., and noting the time the wax takes to melt on each, the following Table

of the conducting powers of different substances has been formed :—

Gold	100	Tin	30·38
Silver.........	97·3	Lead.........	17·96
Copper	89·82	Marble......	2·34
Iron	37·41	Porcelain...	1·22
Zinc	36·37	Brick Earth	1·13

25. Good and bad conductors.—Glass and wood are bad conductors of heat ; thus we may burn a piece of wood in the flame of a lamp, and not feel in the hand any inconvenience from the heat, even at a short distance from the flame. The extremity of a glass tube, held with the hand at one inch from the flame, may be fused with a blowpipe and no unpleasant heat will be experienced ; whereas an iron wire in the same flame will be with difficulty heated to redness, from the heat passing through its whole length by conduction very rapidly, and certainly it could not be held in the hand. Air is a bad conductor, and hence the double windows in cold climates, which include a layer of air, prevent the conduction of heat from the rooms to the outside, whence it would be dispersed.

In the vegetable kingdom the barks of trees exhibit a structure with difficulty penetrated by heat ; the bark is porous, and, in addition to its non-conducting quality, these pores are filled with air; hence the stem of the tree is little affected either by summer's heat or winter's cold.

Ice is a bad conductor, snow still worse; a coating of ice prevents the cooling of the water underneath, while snow protects the delicate roots of plants from the effect of severe cold.

The boiler of a locomotive steam-engine is covered with wood and felt, to prevent the heat being conducted and radiated away: ice is wrapped in flannel when brought from the ice-house, to prevent the external heat from reaching it.

The crust of the earth is composed of bad conductors of heat; hence although the interior of the earth is well known to be of a much higher temperature than the superficial layers, it communicates little heat to the surface, which is supplied chiefly from the sun. Nor does the heat of summer, still less that of mid-day, affect the temperature of mines or wells even a few feet deep; a thermometer at no great depth below the soil would be beyond the influences which affect the surface. At the Observatory of Paris one has been stationed, for the last seventy years, 91 feet below the foundation of the building, which has indicated an invariable temperature of 53°, two degrees higher than the mean temperature of the air above.

Liquids and gases appear to possess little or no

conductive power; if a tube be filled with water and held slantingly over the flame of a spirit-lamp, the superficial layer of water may be made to boil while the temperature of that in the lower portion of the tube remains unchanged; a piece of ice placed there will not be melted.

26. Convection.—The method by which heat is communicated throughout liquids is called *convection.* Heat must be applied to the lower part of the mass; the lowest particles will expand, and, becoming specifically lighter, will rise, while the colder particles descending from above will supply their place. If this process is remarked when water is boiled in a glass tube in which some powder has been thrown whose specific gravity is the same as that of water, a series of upward and downward currents will be exhibited, which will not cease till the source of heat is removed.

Gases, air of course included, become heated in the same way, though the currents have more of a wavy or cloudy form, unless, as is seldom the case, the air is perfectly still, when smoke will ascend perpendicularly, and fogs and vapours will rise in the same manner.

In the ocean the phenomena of convection are exhibited on a grand scale. In the torrid zone the water of the sea weighs less, quantity by quantity, than at the poles, or near them; therefore the cold water of high latitudes has a tendency to flow towards the equator, and form under-currents throughout the ocean; to supply the equilibrium, the warm waters of the tropical regions, forming the upper current, are carried as far as the arctic oceans, warming the air in their passage, and tempering the rigour of the climates of places situated on the sea-coast. It is invariably found that maritime localities, exposed to the influence of these currents, have a less variable temperature than those inland at the same distance from the equator. This difference is partly owing to the oceanic currents, whereby there is kept up a constant circulation of immense masses of water from the tropical regions to the higher latitudes; thus we find the Gulf-stream flowing from the Gulf of Mexico, with an initial temperature of 84°, along the coasts of North America, and across the Atlantic to the shores of Europe. Water has, as will be shown hereafter, a great capacity for heat, and parts with it slowly; so that the Gulf-stream continues of a temperature far above that of the air, after travelling many degrees north: the heating and cooling of the water proceeds very deliberately, and the air partakes in a measure of the uniformity of temperature which belongs to the surface of the ocean; thus the extremes of temperature are much less at places near the ocean than at those at a distance from it.

An illustration taken from two places in the same latitude, but one on the sea-coast and the other inland, Penzance, and Barnaul at the foot of the Altai in Siberia, will tend to impress the mind with this truth :—

	Winter temperature.	Summer temperature.	Difference.
Penzance	44·6	60·4	15·8
Barnaul	6·6	61·9	55·3

27. Laws of heat : *Radiation and Absorption.*—Besides the motion of heat from particle to particle of a body, there is another mode in which diffusion of it takes place; that is by radiation. Any body heated higher than the surrounding medium sends forth rays of heat from its surface. It is thus that we receive heat from the sun, and not by conduction through the air—else it would be as hot, or nearly so, in the shade as in the sun.

The radiating power of a body depends greatly on the character of its surface; smooth or polished radiate more slowly than rough surfaces; dark surfaces more quickly than light.

Absorption of heat radiated from another body takes place in different degrees with different kinds of matter, the power of absorption being directly proportioned to that of radiation.

28. **Laws of heat**: *Reflexion.*—Heat is reflected from surfaces, obeying laws similar to those of reflected light; bright polished metals are powerful reflectors of heat; a red-hot ball in the focus of a parabolic mirror will radiate heat which will be reflected from the curved surface in parallel lines; these being received on the surface of another parabolic mirror will be reflected to its focus, and a powerful heat sufficient to inflame gunpowder or phosphorus will be felt at that point.

A thermometer hung in the sunshine against a wall, or where the rays of the sun are reflected on it, will show a very much higher temperature than one in the sun away from all reflected heat. Such a thermometer not only receives the direct rays, but heat from those which are reflected; and thus the temperature it shows is the result of the combination of direct and reflected heat, and therefore not to be trusted as a fair record of the heating power of the solar rays.

Those bodies reflect heat most perfectly that absorb it least.

29. **Latent heat.**—A very important condition of heat is that called latent. Latent heat is so named because it does not affect the sense of touch or the thermometer; it may be thus illustrated.

If a vessel be filled with ice in a melting state

and subjected to the application of heat, a thermometer placed in it will not show a higher temperature than 32° till the whole of the ice has melted; not till then will the temperature of the water begin to rise. The heat thus absorbed by the ice in passing from a solid to a liquid state has therefore become latent. Heat thus becoming latent on liquefaction will account for the chilling effect of the air during a thaw, for the ice and snow absorb a considerable quantity of heat in the process of melting, which is borrowed from the surrounding air, so that the latter is continually kept down to near the freezing-point of water.

The temperature of cold water, after it has been placed over a fire, will continue to rise until it has attained to 212°, when ebullition takes place and the water passes with violent action into steam: the effect of a continuance or increase of heat will not be that of raising the temperature, which in an open vessel is impossible, but the additional heat will be employed in converting the liquid water into steam.

When matter passes from a liquid to a gaseous state, heat is absorbed and becomes latent: in the process of freezing carbonic acid gas, the gas is first liquefied by pressure; on being allowed to rush out of the vessel into the air it immediately recovers its gaseous state, but for this purpose the portion first liberated demands and obtains from that which remains so much heat, that what is left behind is solidified, the solid carbonic acid presenting an appearance like flakes of snow.

On the contrary, when a vapour becomes liquid the latent heat is given forth and becomes sensible. It requires a much greater amount of cold water to condense a given weight of steam at a temperature of 212°, into water whose temperature shall be 100°, than to cool down a similar weight of boiling water to that temperature; hence, in condensing engines, an immense quantity of cold water is requisite to preserve a sufficient vacuum in the cylinder. It is usual in some manufactories to raise a large quantity of cold water to the boiling-point in a few minutes by passing steam into it; on condensation the steam gives out its latent heat, which the water absorbs.

The three conditions in which matter exists, of solid, liquid and aëriform, differ apparently only in this, that in the one case there is sufficient heat to sustain a liquid or gaseous form, which a solid fails to possess. Water becomes solid at 32°; and even mercury, usually met with as a fluid, will solidify when the temperature falls to −38°·2.

30. Evaporation.—For liquids to assume the gaseous form ebullition is not necessary: this is a violent process of evaporation, and takes place only when the elastic force of the vapour of the liquid, which has been heated up to a certain fixed temperature,—in the case of water 212°, at which point steam is rapidly generated—balances the pressure of the atmosphere. Evaporation from the surface of water proceeds at all temperatures, and goes on gradually and insensibly; the particles of water rise in the air and are mixed with it, and, unless they exist in large amount, are invisible; they do, however, always exist there, and, when condensed, fall in refreshing showers or in storms of rain and hail. Whenever evaporation takes place, heat must be borrowed from contiguous substances to supply the amount which becomes latent in the conversion of the liquid into vapour; hence it is always a cooling process. From the same cause, ether, which evaporates rapidly at a low temperature, applied to the surface of the skin, produces a sensation of cold; and the evaporation of water from the surface of a porous jar cools its contents, as is beneficially experienced in hot climates; in this case the water in the vessel exudes through the pores and forms on the outside a kind of dew, which is taken up by the hot air very rapidly.

The insensible evaporation of water, and the combining of its vapour with air, may be proved by exposing a shallow vessel out of doors full of water; in a few days it will be found emptied of its contents. If the surface of the vessel be of a given area, say one square foot, and the water be weighed from time to time, the loss of weight will give the rate of evaporation, for the temperature and locality, due to one square foot of surface. The process of evaporation is always going on on a large scale over the surface of oceans, seas, rivers, and lakes; and the amount of water thus taken up into the air is enormous.

31. Specific heat.—Some substances have a greater amount of absolute heat than others, although their temperature may be expressed by the same degree of the thermometer: if two glass tubes, in every way alike, one containing water and the other mercury, be subjected to the same degree of heat by being plunged into a vessel of hot water, the mercury will attain the heat of the water in half the time which the water will take; and on removing the tubes and allowing them to cool, the mercury will only take half the time to recover the initial temperature which the water will: this effect arises from the mercury absorbing less heat than the water does in being raised to a similar temperature; as the fact is usually expressed, the water, compared with mercury, has twice the capacity for heat.

In comparing the capacity for heat of different

bodies, it is usual to take equal weights of each, rather than equal volumes: a pound of distilled water takes a certain amount of heat to raise it a certain number of degrees; to this standard other substances are reduced; if a pound of mercury requires ·033 of the same amount of heat to arrive at the same temperature, then the specific heat of water is said to be to that of mercury as 1000 to 33. M. Regnault has conducted a series of very careful experiments to determine the specific heat of various substances, of which the following are some of the results:—

Substances.	Specific heat of equal weights.
Water.........	1000
Ice	513
Iron............	113·8
Copper	95·15
Zinc............	95·55
Glass	198
Mercury......	33·32
Lead	31·4
Air	3705

It will appear from this Table that the capacity of water for heat is very great; hence, as it forms so large a portion of the envelope of the globe, it becomes a great reservoir of heat, and performs an important office in equalizing the temperature of the earth, as we have before seen.

Mercury has little capacity for heat, and therefore is very readily operated upon by the slightest change of temperature. Were there no other reason why it should be adopted for thermometers, this would be sufficient, as no other liquid would exhibit such sensibility.

The air, it will be seen, has a very considerable capacity for heat, varying, however, with its density. Great difficulties and some uncertainty still attend the determination of the specific heat of all gases; it would appear, however, that air cools by expansion to the amount of 40° or 50° in doubling its volume; and that, on compression into half its volume, its heat is raised to that amount. The match syringe is a small cylinder of brass with a tightly fitting piston; on suddenly driving this down to the bottom of the cylinder and compressing the air within into a very small space, sufficient heat is evolved to light an inflammable substance, such as amadou or German tinder. We shall, in the progress of this work, have occasion to refer to the decrease of temperature in the higher regions of the air, as dependent on the circumstance of the capacity for heat increasing with the rarefaction of the upper strata of the atmosphere.

32. Coldness of the upper regions of the air.— Near the surface of the earth the temperature of the atmosphere is observed to diminish 1° in every

352 feet of ascent, but there is reason to suppose that in the higher regions this ratio does not exist: the observations at great altitudes have not been conducted in sufficient amount to enable us to arrive at anything like certainty on this point.

33. **Colour of the air.**—The colour of the atmosphere is, according to Sir David Brewster, due to light which has suffered polarization; the red and yellow rays are absorbed, and hence the beautiful azure of the vault of heaven, which deepens the higher we ascend. The colours which accompany the setting sun are due to the vapours through which his rays reach us when he is near the horizon; the aqueous vapour absorbs the blue rays, and admits the passage of the red and yellow, the blending of which two colours produces the varied tints which surround him.

PART II.

OBSERVATION AND DEDUCTIONS.

34. Arrangement of the subject. — Having cleared our way by establishing general principles, we are prepared to enter upon the explanation of the construction of meteorological instruments, the mode of observation, and the deductions which may be drawn from the observations themselves. The subjects may be treated in the following order:

1. The Thermometric ⎫
2. The Hygrometric ⎬ condition of the air.
3. The Barometric ⎪
4. The Electric ⎭

Under each head will be laid down the principles of construction and the use of meteorological instruments, the cautions to be applied in making observations with them, and the atmospheric laws which may be fairly deduced.

1. *The Thermometric Condition of the Air.*

35. The Thermometer. — The importance of having at hand, at all times, the means of determining the temperature of the air, on which all nations shall be agreed, is self-evident; without this there could be no comparison of results; the mercurial thermometer is the instrument adopted for the purpose. A small portion of mercury is enclosed in a glass tube with a narrow bore, one end of which is blown into a small bulb, and then hermetically sealed: its principle of construction is sufficiently simple; for since mercury expands to a great extent by heat, the amount of this expansion over that of the glass tube, which is slight, affords us the means of estimating the degree of heat due to the surrounding medium. The apparent dilatation of mercury in a glass tube between 32° and 212° (the freezing and the boiling points of water) is $\frac{1}{64\cdot8}$ of its volume; and its true dilatation within the same limits is $\frac{1}{55\cdot5}$. The construction of a thermometer whose indications may be relied upon with certainty in small variations of temperature, is attended with practical difficulties, as will be at once apprehended from the following description of the method of constructing a "standard thermometer;" that is, an instrument whose scale has been divided independently, or without comparison with others, and whose readings ought to be correct throughout the whole extent, within a very small fraction of a degree.

36. Construction of a standard thermometer.—The construction of such a measurer of temperature will embrace three operations:—

α. The choice of a suitable tube.

β. The filling and hermetically sealing of it.

γ. The adaptation of the scale.

α. The tube adapted for a standard thermometer must have either a cylindrical bore, or one whose section is an ellipse, the longer diameter of which must be presented to the eye in reading off, the shorter must be almost evanescent; in this latter construction, the exact point at which the mercury intercepts the scale is very distinct; more so than in a tube of greater capacity whose section is a circle. As the glass, as well as the mercury, expands by heat, the rise of the mercury will be the difference of the expansions of the two in the line of the scale. Mr. Sheepshanks, who gave great attention to the construction of perfect thermometers, found that those with round bores were far more nearly true as regarded uniformity than those with flat bores; but his method of constructing the scale rendered him altogether independent of the workman's skill beyond what may be ordinarily attained.

If the section of the bore of the tube be exactly the same throughout its whole length, the tube is so far perfect; it may be thus tested. Let a small portion of mercury be admitted into the tube, which shall occupy a space (a few degrees) throughout which no error will be likely to arise from any deviation from uniformity if it exists; measure accurately the length of the mercury in several parts of the tube; if the length be invariable, the bore of the tube is the same throughout; if the differences in the measures are variable and very evident, the tube must be rejected altogether, or very nice measurements and calculations entered upon, which need not be described in this place. The subject of the calibration of the tube has been fully discussed by Mr. Welsh, in a report presented to the Royal Society, May 1852, "On the general process adopted in graduating and comparing the Standard Meteorological Instruments for the Kew Observatory."

The bulbs should be neither very small nor very large; if too small, the amount of mercury will be so diminished that the space allotted to a degree will be too much reduced in size for very near readings; if too large, the mercury will take some considerable time to be heated throughout, and will not indicate sudden changes of temperature. Mr. Sheepshanks approves of bulbs 3 or 4 tenths of an inch in diameter. The bulb is never blown by the breath, lest moisture should find its way into the tube; but air is urged in by means of an elastic ball, which is compressed at the instant that the closed end of the tube is heated to the point of fusion.

β. The mercury, which should have been previously boiled to separate it from moisture and air, is introduced into the bulb as follows:—

The extremity of the tube opposite to the bulb is expanded into the shape of a funnel, *a*, Plate I. fig. 4, for the purpose of receiving a quantity of mercury (some workmen only form a temporary funnel of paper); the air in the bulb and tube is driven out by the heat of a spirit-lamp, and the mercury takes its place when the lamp is removed and the tube cools; the mercury is then gently boiled in the tube for some time. In sealing the tube, the mercury, by expansion, is made to fill it entirely; and, at the instant of its arrival at the upper extremity of the tube, the flame of a blowpipe is made to melt the glass and hermetically close it; so delicately is this process performed, that not a particle of air is allowed to remain within. The test of the absence of air is a simple one. After the tube has cooled, incline it, and the mercury will flow out of the bulb down to the other extremity if no air be present; but if any remain, it will form a cushion against which the mercury will impinge and never reach the end.

An experienced workman will readily judge how

much mercury is necessary to be introduced into a tube of a particular range; he may do so by plunging the tube into water of different temperatures, and, observing the length of the column between the two temperatures, he will manage to insert the necessary quantity.

γ. To adapt a correct scale to the thermometer is a matter of the utmost consequence. The civilized world has agreed on fixing two points on the scale, viz. that to which the mercury rises when the instrument is plunged in boiling water, and that to which it descends when placed in melting ice. The space between these points is not however divided by all nations into the same number of equal parts, and hence, as we shall see, the degrees of one scale must be converted into the corresponding number of another, before temperatures registered under one method can be compared with those of a different one.

37. **The zero-point.**—The first point to be determined is the freezing-point of water, which is the same under all temperatures and every variety of barometric pressure, provided the water is free from salts of every kind. The thermometer tube —not the bulb only—is plunged into melting ice, and a mark is made across the glass of the tube at the point to which the mercury sinks, or withdraws, towards the bulb; this is the zero of every

scale except Fahrenheit's, the one in general use in England.

It is a remarkable fact that this zero-point is not permanent; when thermometers have been made for some time, a very perceptible difference is found between the point to which the mercury descends when plunged in melting ice, and the original zero of the scale; hence the necessity of determining it anew from time to time, and applying the difference as an index correction. A tube intended for a standard thermometer is allowed to remain, after it has been filled, for at least six months before the freezing-point is marked on it, in order that the glass, which has been heated to the boiling-point of mercury, may recover its normal state. If the zero were marked at an earlier period of its existence, there would be a rise during the first year of considerable amount.

Certain peculiarities have been remarked in the construction of thermometers by Mr. Welsh, which he has thus described :—"If a thermometer, after having been for some weeks exposed to the ordinary temperature of the air, is placed in melting ice, its freezing-point may be, for example, 32·2; if the bulb be then put for two or three minutes into boiling water and soon afterwards placed in ice, the reading will have fallen to 32°: if in a day or two it is again placed in ice, the freezing-point will have risen a little—about 0°·1; and if tried again after two or three weeks, the freezing-point will be found to have acquired exactly the original position of 32·2." It would appear that the glass, after a considerable change of temperature, requires a long time to return to its normal dimensions.

Mr. Sheepshanks, who did not desist from his labours in the structure of the thermometers used in connexion with the experiments to determine the length of the National Standard Yard till he felt assured of being able to reach the second decimal of a degree, used ice found in tubs of rain-water, in which the thermometer was placed horizontally, and the intersection of the mercurial column with the tube was read off with a vertical telescope carrying a wire moved by a micrometer-screw; he obtained identical results when he made use of newly-fallen snow; and he recommends the determination of the zero to be performed in winter, when the air around is but little above the freezing-point, and the ice or snow melts but slowly.

38. **The boiling-point.**—The other constant—the boiling-point—is not so easily ascertained, seeing that the temperature of boiling water will differ with the pressure of the atmosphere.

In the division of the scale adopted by the

Commissioners appointed by Government to construct Standard Weights and Measures, and also by the Kew Committee of the British Association, the boiling-point, 212°, is made to represent the temperature of steam under Laplace's Standard Atmospheric Pressure, which corresponds to the following number of inches in the barometric reading, reduced to 32° Fahrenheit,

$$29{\cdot}9218 + 0{\cdot}0766 \times \text{cosine (2 Lat.)} + (0{\cdot}00000179 \times \text{height in feet above the sea-level}).$$

The boiling-point, therefore, will be fixed at 212°, neglecting the small correction for height, when the barometer reading, reduced to 32° Fahr., amounts, at

London, to	29·905	inches.
Dublin, to	29·900	„
Edinburgh, to......	29·893	„

If the reading of the barometer differ from this quantity, and the temperature, at the time of the experiment, be not 32°, an allowance must be made whereby the distance on the scale between the boiling-point and the freezing-point must be increased or diminished.

According to Wollaston, 1° of Fahrenheit will correspond to a difference of 0·589 inch of barometric pressure. Hence if the barometric reading when reduced is one inch below the standard pressure, water will boil at 210°·3 F.; and when the reading is one inch above, the temperature of the boiling-point will be 213°·7.

Mr. Sheepshanks thus describes the boiler he made use of in his experiments :—

"My boiler is made of sheet copper, square above and cylindrical below; the dimensions are, length 24 inches and depth about 6 inches, with flat ends and top. In one end there is a round hole filled with a large cork; through the centre of the cork a small pipe of copper pierces, large enough however to allow the thermometer tube to pass: a bar, stretching across the inside of the boiler at the same height nearly as the centre of the cork, supports the thermometer near the bulb, when the thermometer, cork and all, is inserted pretty tightly in its hole. As much of the tube of the thermometer is exposed as will show the division below the boiling-point, and the joint between the tube and the pipe, which projects a little, is made good with a binder of very thin vulcanized India-rubber. Distilled water is poured into the boiler, but not so much as to touch the thermometer, which is thus boiled in steam. There are some round holes in the flat top which can be closed sufficiently by flat pieces of brass. When the steam rises strongly the flat pieces on the top of the boiler begin to chatter, and it is certain that the necessary heat, at least, is attained. By removing one, two or three of the flat pieces, it will be found that, in a little time, the position of the mercury becomes steady, and is not affected by closing one of the holes or unclosing another. The steadiness of the boiling-point, whether the steam issues languidly or with considerable vehemence, is rather a puzzle to me, but the fact is quite certain."

In Plate I. fig. 5, is represented the cylindrical boiler: *a* is the cork through which the thermometer tube is inserted into the steam without touching the water; *b* is the escape tube; *i* the spirit-lamp; *w* the water used, which must be pure rain or distilled water, for the admixture of any salt would raise the boiling-point.

39. The Scale. Fahrenheit's division.—We have now on the scale two fixed points, and unless we aim at extreme nicety, we have only to divide the space between them into a certain number of equal parts, extending the divisions above the boiling-point and below the freezing, and the scale of the thermometer will be complete. In the thermometer in general use in England, this space is divided into 180 portions called degrees; it derives its name from Gabriel Fahrenheit, a native of Dantzic, who fixed his zero at the point of the lowest cold observed in Iceland, which was supposed to be as low a temperature as was likely to become the subject of philosophical investigation: this zero is 32° below the freezing-point, and 212° below the boiling-point of water. The advantages of this scale, which it possesses above others, are, that the observer, especially in meteorology, is very seldom troubled with *negative* degrees, which do not commence till the extreme cold of 0, or 32° below freezing, has been reached; as the divisions moreover are more numerous than in other methods of division, we do not so frequently find it necessary, in ordinary operations, to use fractions of degrees.

40. Divisions of the Scale according to Celsius and Reaumur.—Celsius, a Swedish professor at Upsal in 1742, proposed to divide the space between the two standard points into 100°, the zero being the freezing-point; temperatures above this are positive quantities, and those below, the scale being continued as far as may be desirable, are negative. A thermometer thus divided has a "Centigrade scale" and is in very general use in France. The principal objection to this division is that, in a record of degrees of natural heat, one column may be embarrassed by + and − degrees.

Reaumur's thermometer has the same space divided into 80°, and also necessitates the use of + and − to mark what are sometimes inaccurately called degrees of heat and degrees of cold; it is in extensive use in Germany and Switzerland, and in Spain.

As meteorological records are kept under all these three systems, it is of some considerable importance to have a formula for readily converting the degrees of one into those of another. Water freezes at 32° of Fahrenheit's scale, and at 0° of the Centigrade and Reaumur's; while the boiling-points are respectively 180, 100, and 80 above that point: hence the number of degrees of Fahrenheit in a given range of temperature are to those of Celsius as 180 : 100, *i.e.* as 18 : 10 or as 9 : 5; and to those of Reaumur as 180 : 80, *i.e.* as 9 : 4; to convert, therefore, Centigrade degrees into those of Fahrenheit, we must multiply them by 9, divide by 5, and add 32°, which is the number of degrees marked in Fahrenheit's scale at the freezing-point, or zero of the others; conversely, to convert degrees Fahr. into degrees Cent., subtract 32, multiply by 5, and divide by 9. By substituting 4 for 5, we obtain the same results on Reaumur's scale.

The following formulæ will meet every case, if the negative sign is used when the reading is below zero:—

$$1.\quad F = \frac{9C}{5} + 32.$$

2. $C=\frac{(F-32)\times 5}{9}.$

3. $F=\frac{9R}{4}+32.$

4. $R=\frac{(F-32)\times 4}{9}.$

5. $C=\frac{5}{4}R.$

6. $R=\frac{4}{5}C.$

Since, however, in reading or writing, these reductions perpetually occur, it is very useful to have at hand such a table as Table I. (Appendix), by which we are spared the labour of reduction, and obtain the corresponding degrees immediately by inspection.

It is seldom that a thermometer used in this country for meteorological observations reads higher than 120°; in such cases, for the thermometer to be valuable, the freezing-point is determined in the same way as for a standard, and every precaution is taken to ensure a tube with an equable bore: an upper reading is fixed on, by plunging it, with a standard, into water of a certain temperature, and the space between the freezing-point and the height to which the mercury rises, the degree of which temperature is shown by the standard, is divided into as many degrees as the temperature is above 32°; the divisions being carried above and below as far as may be required to complete the scale.

The accuracy with which such thermometers are constructed by good workmen is shown by Mr. Glaisher in his "Essay on Radiation"; he says that he received from Messrs. Watkins and Hill upwards of fifty instruments, whose extreme difference of reading from the standard with which they were compared was, in one thermometer a constant quantity of half a degree, in three others a constant quantity of 0°·2 or 0°·3, the remainder being absolutely without error.

41. **Comparison of thermometers.**—If a thermometer has to be compared with a standard for the purpose of ascertaining the differences of its readings throughout the whole range of the scale, the best method is to put them both into water heated up to the highest degree the thermometers read, and record the degree of temperature shown by both every quarter of an hour, taking care to agitate the water from time to time: this record, if the readings differ, will supply a number of index corrections to be applied to the thermometer to bring it up to that of the standard; and it may be presumed, if that be carefully done, that an inferior thermometer may be rendered nearly as useful as a standard, though, it must be confessed,

at some considerable expense of labour. At the present time, however, there is no excuse for using an inferior instrument. The Kew Committee of the British Association have caused to be constructed by Mr. Welsh, at the Kew Observatory, a large number of standard thermometers which are sold to the public at a moderate price; they have all been subjected to the most accurate scrutiny, and the scales are engraved on the tubes, which are of considerable length so that parts of degrees may be estimated with very close precision.

42. **Maximum and Minimum Thermometers.**—To obtain a record of the greatest heat attained during the day, and of the lowest reading of the thermometer during the night, is a matter of much importance in observations on climate, and various means have been adopted to ensure such registration. The thermometers used in this country are Sixe's and Rutherford's, as well as a new maximum and a new minimum thermometer patented by Messrs. Negretti and Zambra which have received the approbation of many practical meteorologists.

43. **Sixe's Register Thermometer.**—Mr. Sixe described his register thermometer originally in the Philosophical Transactions, vol. lxxii. It is, in fact, a spirit of wine thermometer, with a long cylindrical bulb, and a tube bent in the form of a

siphon with parallel legs, and terminating upwards in a small cavity. A portion of the two legs of the siphon tube, from *a* to *b* (Plate II. fig. 1), is filled with mercury, the bulb and the whole of the rest of the tube with spirits of wine; the double column of mercury gives motion to the two indices *c* and *d*, each of which is a piece of iron wire capped with enamel at each end; they would move freely in the tube and rest on the mercury were it not for springs, made of a thread of glass or a hair, which, surrounding them, press against the side of the glass with sufficient power to keep either index stationary in the spot where it is left by the retreat of the mercury. The action of the instrument is as follows: when the increase of temperature expands the spirit in the lengthened bulb G, the mercury in the leg of the siphon, *a*, is depressed, and a corresponding rise takes place in the leg *b*; in its rise the mercury urges before it the index *d*, which, being retained at its highest point by the spring, does not follow the retreat of the mercury, as the temperature decreases and the spirit in G contracts; this shows then the maximum heat in any determined period of time. When the spirit contracts, the mercury descends in the tube *b* on account of the elasticity of some compressed air in the small bulb above, and proportionately rises in *a*, urging before it the index *c*; leaving, on the increase of heat, its lower extremity exactly at the highest point to which the column (the degrees decreasing upward) had risen on that side, pointing out in fact the minimum degree of temperature attained.

To prepare the instrument for future observations, the indices are brought down, by a magnet, to touch the mercurial column. If this instrument be read every twenty-four hours, it will evidently give the greatest and least temperature during the day.

Unfortunately this elegant instrument can hardly be trusted for very nice observations, and is very liable to get out of order. The use of two liquids, both expanding in different degrees, is a defect; and although it may, in part, be remedied by very nice dividing, and comparing the scale with a standard for every 5° or 6°, yet this process would increase the price considerably. The other defect arises from the liability of the springs to get out of order—a glass one by breakage, and the hair by losing its elasticity after long immersion in the spirit; while, from the perpendicular position of the tube, agitation from the wind is very likely to cause the indices to slide down the tube, and thus the observations would be lost. For these reasons Rutherford's maximum and minimum thermometers are preferred for registering the greatest and least degrees of heat; these we shall proceed to describe.

44. **Rutherford's Register Thermometer** (Plate II. fig. 2).—A represents a spirit thermometer, B a mercurial, each furnished with a scale and fixed horizontally on the same plate of box-wood or metal; B contains within it a steel index, *c*, which is urged forward as the mercury expands by heat, and is left to indicate the highest temperature attained when the metal again contracts.

The spirit thermometer, A, contains a glass index, *n*, half an inch long, with a small knob at each end; it lies in the tube and allows the spirit freely to pass it as it expands; when contracted by cold, in consequence of the capillary attraction between the spirit and the glass index, the last film of the column of spirit is sufficient to overcome the slight friction of the index on the inside of the tube and to carry it backwards towards the bulb; it will rest, on the spirit again expanding, at the lowest degree of temperature attained within a given period. After reading off, and to prepare the instruments for future observations, both indices are brought to the extremities—the one of the column of spirit, the other of that of mercury —by gently inclining the plate on which the thermometers are fixed, downward from the horizontal position. This must be done with some care, or the indices will get entangled with the liquids, from which they will be with difficulty extricated. For three years the author has used a register thermometer of this construction without any mishap, though he ruined many before he discovered the careful treatment they required.

45. **Negretti and Zambra's Maximum and Minimum Thermometers.**—The maximum thermometer of Messrs. Negretti and Zambra is strongly recommended: Fig. 3. Plate II. will explain its construction.

The tube is originally straight throughout; in this state a small piece of enamel, *a*, is introduced down it to within a short distance from the bulb; by means of a spirit-lamp the tube is then bent just at the point where the enamel rests, and the heat required for this purpose is sufficient to cause its adhesion to the glass. The enamel does not fill the tube, but allows the mercury to pass freely above it; on the decrease of heat, all that part of the mercurial column which has passed the enamel is left in the tube, while that portion nearer the bulb is separated and withdraws from it, the maximum heat will therefore be shown by the extremity of the detached mercurial column; after this is registered, by depressing the bulb the detached column may be made to reunite with the rest of the mercury, and thus the instrument is prepared for another observation. The advantage of the instrument seems to be, that there is no index to get out of order; the disadvantage, that there is some considerable trouble in accurately determining the corrections to bring its readings in unison with a standard thermometer. The following is the construction of the minimum thermometer of Messrs. Negretti and Zambra (Plate XI. fig. 1) :—*a b* is a long bulb containing much more mercury than is usual in a thermometer, this enables the tube *c d* to be made of a large bore without diminishing the length of a degree on the scale; the tube is prolonged and blown at the end into a bulb, *d e*; a steel needle or index (*f*), pointed at the lower end, lies loose in the tube. The instrument is kept upright, and yet able to be moved to a horizontal position by being hung from a single support; the temperature is reckoned from the upper end of the index when it just floats on the surface of the mercury (as it will do on account of the capillary repulsion); as the mercury sinks in the tube on loss of heat, the needle falls with it, but when it expands it passes along the side of the needle without raising it at all, jambing it indeed against the glass: so the higher end of the index will always denote the lowest temperature reached.

To free the index from the mercury so as to set

the instrument for another observation, the lower part must be raised till the mercury that is in the tube runs down into the small bulb at the end, leaving the index free; this is then led down to the same place by a magnet, and held there while the thermometer is being restored to its former position, on which the mercury will fall back to its proper place, and the index may then be gradually lowered till it just touches the surface. Great care must be taken not to let the needle fall on to the mercury from a height, else it will plunge in and cause the reading of the thermometer to be too low. If before an observation is recorded the mercury has risen above the top of the index, and the degree this points to cannot well be observed, the magnet should be applied so as to hold the needle in its place (for the general position of it is told by the slight displacement of the mercury), and the bulb elevated as before directed till the index is free from quicksilver, when the exact reading can be taken, and the instrument readjusted as usual.

A great advantage of this form of thermometer is that it is not at all likely to get out of order, from the absence of air and from the large bore of the tube. A disadvantage is, that its reading may not always be correct to a decimal of a degree, because the index, unless carefully managed, may sink too low in the mercury.

46. Hick's Maximum and Minimum Thermometer.—With this newly-invented thermometer, which is made by Mr. Casella of Hatton Garden, the extremes of heat and cold are registered by one instrument, and by the expansion and contraction of one fluid, mercury, though spirit is used for carrying one of the indices. Its construction will be understood from the drawing of it in Plate XI. (fig. 3.). It is a thermometer with its tube *a b c* bent at right angles and ending in a small bulb at *c*; the mercury of the bulb extends some way into the upright part of the tube and moves an index *d*, like those of Sixe's thermometer, which is left at the highest point reached by the mercury, and therefore registers the maximum heat, read off from the scale *f g*; from the top of the mercury, as far as *e*, the tube contains spirits of wine, which, on the mercury contracting, also recedes (being pressed back by the air at *c*), carrying with it the index *e*, which will be left to mark the lowest reading, the scale *h i* denoting it. This upper scale is graduated in such a manner as to compensate for the expansion and contraction of the spirit, which however affect the spaces for the degrees to a very small extent.

In a maximum thermometer devised by Professor Phillips, a portion of the mercurial column was separated from the rest by the intervention of a small bubble of air; this portion, as in the case of Negretti's, remains in the tube on the contraction of the mercury in the bulb, and thus serves to mark the highest reading attained; by the inclination of the tube it is then brought to its original position preparatory to another observation. This thermometer does not seem to have received the attention it deserves; its construction is simple and its indications sure, as those can testify who have used it.

47. Observations to determine temperature. *Diurnal Range.*—Provided with a good thermometer for general purposes, and with others for recording the highest and lowest temperatures occurring within a definite period, we are in a position to take observations on the temperature of any place which shall be of some value in a scientific point of view.

One of the first desiderata will be the mean temperature of the place.

We cannot arrive at this without a series of observations extending over at least three years; we might, if time and means allowed, advantageously begin with twelve observations in the course of the day, taken at the even hours, throughout the year; from which we could deduce the mean temperature of each day by dividing the sum of the temperatures recorded by 12; of the month, by dividing the sum of the mean daily temperatures by the number of days it contained; and of the year, by dividing the sum of the mean monthly temperatures by 12, the number of months in a year. As however few observers are in circumstances favourable to the undertaking of so elaborate a series, we shall, for general advantage, show that we have the power of eliciting the mean temperature with far less expenditure of time and labour.

For many years, at the Royal Observatory, Greenwich, the thermometer (as well as other meteorological instruments) was registered every two hours throughout the 24, under Mr. Glaisher, the Superintendent of the Meteorological department, and the results he has deduced from the series of observations, extending from 1841 to 1845 inclusive, are published in the 'Philosophical Transactions,' Part 1, for 1848. Mr. Glaisher has since been able to extend his investigations by reducing observations antecedent to the elaborate work performed at Greenwich, and thus he has determined the average mean temperature of each day throughout 38 years, and from this has obtained much valuable knowledge of the laws of the daily increase and decrease of temperature. Some of the results of his labours we shall now place before the reader.

The monthly mean of the readings of a thermometer at a particular hour differs from the monthly mean temperature by a quantity which is constant for each particular month; by applying, therefore, this quantity as a correction, we get a true monthly mean from our single daily observation. The numbers to be applied as corrections, for the several months, were thus obtained. Two-hourly observations at Greenwich, commencing at noon, from January 1, 1841 to December 31, 1845, were the data from which was calculated the mean temperature of every day in the year. The mean of the daily temperatures for the month gave the mean temperature for each month. Combining then the whole series of observations taken at any particular hour—say at noon—during each similar month of the five years, the difference between the temperature due to that hour (which was found by dividing the sum of all the noon observations for the month, during the five years, by their number) and the mean temperature for the month was noted. This process was repeated for every alternate hour in the day; and thus the excess of the mean value at each even hour in the day for every month, above the mean value for the month, or else the amount by which it fell short of it, was arrived at.

The accordance in the results of observations taken at the same hour in the same month in the different years, was found to be very close and satisfactory—the small discrepancies disappeared when the whole series was combined.

48. Projection of the same.—To illustrate the meaning of what we have just stated, we must refer to Plate IV. fig. 2; along the horizontal line are set off the hours for *one day*; at each hour for the hottest and coldest months of the year, July and January, ordinates are taken equal to the differences between the mean of the observations due to that hour, and the mean temperature for the month: a curve which shall pass through the extremities of these ordinates will show the rise and fall of the temperature throughout the period of twenty-four hours for either month—this will be the normal diurnal curve for the month; and although from various causes the heat at each hour may, on any day, be greater or less than that indicated by the point where this curve cuts an ordinate, yet, in the run of a month, the irregularities will in general be neutralized; those above the mean being compensated by those below.

These remarks will explain the object of the

following Table, which shows the "corrections to be applied to the monthly mean reading of a thermometer, placed at the height of 4 feet above the soil, with its bulb freely exposed to the air, but in other respects protected from the influence of radiation and rain, at any hour, to deduce the true mean temperature of the air for the month from the observations taken at that hour."

Local mean time.	Jan.	Feb.	March.	April.	May.	June.
h	°	°	°	°	°	°
12 Midn.	+0·9	+1·6	+2·9	+4·8	+5·4	+6·2
1 A.M.	+1·0	+1·8	+3·0	+5·2	+6·0	+7·1
2	+1·2	+2·0	+3·3	+5·7	+6·4	+8·0
3	+1·3	+2·1	+3·6	+6·2	+6·7	+8·7
4	+1·6	+2·3	+3·9	+6·6	+6·7	+9·3
5	+1·8	+2·2	+4·0	+6·7	+6·3	+8·8
6	+1·9	+2·3	+3·9	+6·0	+4·8	+6·4
7	+1·9	+2·1	+3·6	+4·3	+2·6	+3·0
8	+1·5	+1·6	+2·5	+2·0	+0·5	0·0
9	+1·0	+0·7	+0·2	−0·9	−2·0	−2·5
10	+0·2	−0·5	−1·9	−3·2	−4·0	−4·5
11 A.M.	−1·3	−2·1	−3·5	−5·3	−5·5	−5·8
12 Noon.	−2·3	−3·2	−5·0	−6·8	−6·7	−7·3
1 P.M.	−2·9	−3·9	−5·8	−7·9	−7·5	−8·1
2	−3·0	−3·9	−5·8	−8·2	−7·7	−8·6
3	−2·5	−3·6	−5·5	−7·7	−7·3	−8·4
4	−1·9	−2·8	−4·5	−6·7	−6·1	−7·4
5	−1·1	−1·6	−3·3	−5·4	−4·8	−6·1
6	−0·6	−0·6	−1·8	−3·5	−3·0	−4·5
7	−0·3	+0·3	−0·4	−1·1	−1·0	−2·4
8	+0·1	+0·6	+0·9	+0·7	+0·9	0·0
9	+0·4	+1·0	+1·7	+2·0	+2·3	+1·8
10	+0·6	+1·3	+2·3	+3·2	+3·5	+3·6
11 P.M.	+0·7	+1·5	+2·6	+4·1	+4·5	+5·0

Table continued.

Local mean time.	July.	Aug.	Sept.	Oct.	Nov.	Dec.
h	°	°	°	°	°	°
12 Midn.	+5·0	+5·1	+4·0	+2·9	+1·7	+0·9
1 A.M.	+5·5	+5·5	+4·5	+3·0	+1·8	+1·0
2	+6·0	+6·0	+5·5	+3·4	+2·0	+1·2
3	+6·4	+6·3	+6·4	+3·6	+2·0	+1·3
4	+6·6	+6·5	+6·6	+3·8	+2·1	+1·4
5	+6·2	+6·5	+6·2	+3·8	+2·0	+1·4
6	+4·5	+5·5	+5·3	+3·5	+1·9	+1·4
7	+2·5	+3·3	+4·0	+2·8	+1·7	+1·5
8	0·0	+0·9	+2·1	+1·6	+1·0	+1·3
9	−2·0	−1·6	−0·4	0·0	+0·4	+0·9
10	−4·0	−3·5	−3·0	−2·0	−0·6	0·0
11 A.M.	−5·4	−5·4	−5·0	−3·8	−2·0	−1·3
12 Noon.	−6·4	−6·5	−6·4	−5·1	−3·1	−2·1
1 P.M.	−6·7	−7·5	−7·1	−5·5	−3·5	−2·4
2	−6·7	−7·7	−7·1	−4·9	−3·6	−2·3
3	−6·5	−7·0	−6·6	−3·7	−3·0	−1·9
4	−5·8	−5·5	−5·5	−2·8	−2·1	−1·3
5	−4·9	−3·6	−4·2	−1·7	−1·2	−0·8
6	−3·5	−2·0	−2·5	−0·8	−0·4	−0·4
7	−1·5	−0·5	−0·6	0·0	+0·1	−0·1
8	+0·3	+1·0	+1·0	+0·7	+0·6	+0·2
9	+1·9	+2·4	+1·8	+1·3	+1·0	+0·4
10	+3·3	+3·3	+2·7	+1·9	+1·3	+0·5
11 P.M.	+4·2	+4·3	+3·4	+2·4	+1·5	+0·8

By comparing this Table with the projection, Plate IV. fig. 2, its use will be immediately apprehended. Let us suppose that observations have been taken at the noon of every day in July, and that the mean of these is 69°; from the projection, we see that the normal curve at noon rises between 6° and 7° above the mean temperature, and the table gives us the exact quantity, namely 6°·4, which must be applied with a negative sign, to obtain the mean temperature for that month; therefore we subtract this amount, and the remainder will be the element sought, viz. 69—6·4, or 62·6.

49. Hours in the day when the thermometer shows the mean monthly temperatures at Greenwich.—The January curve will be observed to cross the line of mean temperature twice in the twenty-four hours, viz. at 10 A.M. and 8 P.M.

The July curve crosses its line of mean temperature at 8 A.M. and at $8^h\ 5^m$ P.M.

In like manner it was found that at certain times of the day throughout every separate month, the temperature of the air was at its mean value; these times are as follows in the several months:—

	h m		h m
In January ...at	10 0 A.M.	and again at	8 0 P.M.
In February „	9 30 A.M.	„	6 40 P.M.
In March „	9 10 A.M.	„	7 20 P.M.
In April „	8 40 A.M.	„	7 0 P.M.
In May „	8 25 A.M.	„	7 30 P.M.
In June „	8 0 A.M.	„	8 0 P.M.
In July „	8 0 A.M.	„	8 5 P.M.
In August „	8 20 A.M.	„	7 20 P.M.
In September „	8 55 A.M.	„	7 20 P.M.
In October „	9 0 A.M.	„	7 0 P.M.
In November „	9 25 A.M.	„	6 45 P.M.
In December „	10 0 A.M.	„	7 20 P.M.

In a very elaborate treatise 'Sur le Climat de la Belgique,' by M. Quetelet, founded on observations taken at Brussels during the ten years 1833–1842, may be found discussions of a similar character to those of the Greenwich observations. The following Table of the times when the mean temperature is reached in each month at Brussels is extracted from it:—

	h m		h m
In January...at	9 30 A.M.	and again at	6 40 P.M.
In February ,,	9 42 A.M.	,,	7 12 P.M.
In March ,,	9 24 A.M.	,,	8 0 P.M.
In April ,,	8 40 A.M.	,,	8 0 P.M.
In May ,,	8 30 A.M.	,,	7 58 P.M.
In June ,,	8 24 A.M.	,,	8 5 P.M.
In July ,,	8 12 A.M.	,,	8 6 P.M.
In August ,,	8 42 A.M.	,,	7 57 P.M.
In September ,,	8 48 A.M.	,,	7 35 P.M.
In October ,,	9 6 A.M.	,,	7 9 P.M.
In November ,,	9 15 A.M.	,,	7 12 P.M.
In December ,,	9 36 A.M.	,,	8 0 P.M.

It is generally found that the greatest heat of the day is attained about 2 P.M., and the least a short time before sunrise in every month of the year, at Greenwich: on this point M. Quetelet says, referring to Brussels, "Ces deux termes critiques varient, quand on considère séparément les différents mois. La plus haute température, pendant les jours de janvier, tombe à 1^h 34^m après midi, et s'éloigne d'autant plus de ce point qu'on se rapproche davantage de l'été. Dans cette saison, c'est trois heures qui est l'époque la plus chaude de la journée. La température la plus froide du jour arrive vers 6 heures du matin en hiver; et, en été, vers 3 heures et $\frac{1}{4}$."

As it is instructive to compare phenomena observed under different circumstances and in various localities, the corrections for diurnal range for January and July at Brussels are here given; they are deduced from observations by M. Quetelet, extending through 1841–1844, Centigrade degrees being expressed in those of Fahrenheit. These may be projected and compared with the normal diurnal curve at Greenwich (Plate IV. fig. 2).

Hours.	Jan.	July.	Hours.	Jan.	July.
Noon.	−1·8	−4·5	Midnight.	+1·6	+4·9
P.M. 2	−2·2	−5·6	A.M. 2	+1·8	+3·3
4	−1·3	−5·6	4	+1·8	+6·5
6	+ ·2	−4·5	6	+1·8	+4·3
8	+ ·7	0·	8	+1·6	+ ·5
9	+ ·9	+1·8	9	+ ·9	−1·3
10	+1·	+3·1	10	− ·2	−2·7

50. Proper times of observation.—To determine the mean temperature of the air, it might be thought advisable to take an observation at one of the two periods of the day when the mean temperature is reached; but it must be borne in mind, that at these times the changes of temperature are rapid, as may be seen by the sudden ascent and descent of the branches of the curve (Plate IV. fig. 2), and, consequently, if the observation is made a little too soon or a little too late, considerable errors might be committed; observations at these times, therefore, unless they are made very accurately with regard to time, are not worthy of implicit confidence.

The general hours of observation adopted by those whose time is occupied by other than scientific matters are either 9 A.M. only, or 9 A.M., 3 P.M., and 9 P.M.: observers commencing a series would do well to follow plans already in use, whereby their registers will be more readily comparable with those in progress or existence.

51. Glaisher's factors.—From inspection of the projected curve of mean diurnal range for the month, it will be seen to assume the form of a single progression, having one ascending branch and one descending: the maximum heat occurs early in the afternoon, generally at 2 P.M., and the minimum at about sunrise. For other localities differing in latitude from Greenwich, this curve may not be of the same form; the range of temperature may be greater or less; if so, the apex of the curve will be more or less pointed, and the return from the lowest point more or less sudden; but, amidst all the varieties which may occur, the maximum and minimum heat, or the turning-points of the curve, will most probably occur at nearly the same local time, if the latitude be not very different. Hence it may be concluded, that the amount of the correction applicable to observations taken at any hour at any place, is the same part of the whole monthly mean daily range at that place, as the correction at Greenwich is of the monthly mean daily range at Greenwich. Now this element is found by adding up the readings of the maximum thermometer daily and taking the mean; doing the same with the minimum; then the mean of the maxima, minus the mean of the minima, will be the mean daily range. The following extracts from the Tables already quoted, founded on the above assumption, will give certain factors, which, multiplied into the mean daily range, will give the correction to be applied to hourly observations to obtain the mean temperature for the month, for any other place beside Greenwich.

Factors to be multiplied into the mean daily range of the thermometer to deduce the correction to be applied to the monthly mean reading at the hours 9 A.M., 3 P.M., and 9 P.M., to determine the true mean temperature of the air for the month:

Hours.	Jan.	Feb.	March.	April.	May.	June.
9 A.M.	+·123	+·074	+·015	−·053	−·112	−·128
3 P.M.	−·309	−·383	−·414	−·458	−·410	−·431
9 P.M.	+·049	+·106	+·128	+·119	+·130	+·092
Hours.	**July.**	**August.**	**Sept.**	**Oct.**	**Nov.**	**Dec.**
9 A.M.	−·116	−·094	−·025	0·	+·043	+·106
3 P.M.	−·376	−·410	−·408	−·340	−·316	−·224
9 P.M.	+·110	+·140	+·111	+·120	+·106	+·048

These factors are formed from the Table, § 48, by dividing the correction for the hour by the mean daily range, and it is considered that just such a portion as the hourly correction is of the mean daily range at Greenwich, such portion will the correction for any other place be, in which the range of the thermometer during the twenty-four hours may be greater or less, of the mean daily range at that place.

52. Example of their application.—The application of them will be very evident from an example which I shall extract from my own observations during the month of January 1849, at Southampton; taken at 9 A.M., 3 P.M., and 9 P.M.

Mean of the 9 A.M. readings from the month ... 40·7
" " 3 P.M. " " ... 44·1
" " 9 P.M. " " ... 41·3

Mean of the maxima 45·3; of the minima 37·4.

Mean daily range 45·3 − 37·4 = 7·9.

Correction for 9 A.M.

$7{\cdot}9 \times +{\cdot}123 = +{\cdot}97$; $40{\cdot}7 + {\cdot}97 = 41{\cdot}67$ mean temperature for the month from the 9 A.M. observations.

Correction for 3 P.M.

$7{\cdot}9 \times -{\cdot}309 = -2{\cdot}44$; $44{\cdot}1 - 2{\cdot}44 = 41{\cdot}66$ mean temperature from the 3 P.M. observations.

Correction for 9 P.M.

$7{\cdot}9 \times +{\cdot}049 = +{\cdot}39$; $41{\cdot}3 + {\cdot}39 = 41{\cdot}69$ mean temperature from the 9 P.M. observations.

Referring to the Table in § 48, let us deduce the mean temperature by the application of the Greenwich corrections:

	9 A.M.	3 P.M.	9 P.M.
	40·7	44·1	41·3
Cor.	+ 1·	− 2·5	+ ·4
M. T.	41·7	41·6	41·7

The similarity in these results very satisfactorily proves the extended utility of the Greenwich factors, and shows that the mean daily range of the thermometer at Southampton differs very little from that at Greenwich, for this month at least. Although the month was not chosen with design, yet I am bound to say that seldom does the mean temperature result so identical from the three daily observations as in the example above worked out.

53. **Monthly mean temperature from maxima and minima.**—The relation between the temperature taken at any hour of the day throughout the

month and the mean temperature for the month having been thus satisfactorily established, it was found that the method of determining this latter element, which had been generally in use, viz. by deriving the mean temperature from the arithmetical mean of the maxima and minima, was to some extent erroneous, and that the mean of the maxima and minima needed a correction corresponding to the time of the year, before the mean temperature could be deduced; that only in the month of December was no correction requisite; that such a mean temperature would be too great during every other month of the year by the following amounts, which must, therefore, be subtracted from the means of the maxima and minima readings before a true result could be obtained:—

January...	0°·2	July.........	1°·9
February	0·4	August ...	1·7
March ...	1·0	September .	1·3
April......	1·5	October ...	1·0
May	1·7	November .	0·4
June......	1·8	December..	0·0

These quantities, strictly speaking, are only adapted to the middle day of each month; if it be required to obtain the temperature for a particular day, and not the monthly mean, the following plan may be adopted.

Let the correction for any month, say June, be represented by b; for July by c; the difference between these two divided by 30, the number of days between the middle days of the two months, and multiplied by (a) the number of days between June 15 and the particular date in question, will give a number which must be added to the June correction; thus,

$$\frac{\text{max.}+\text{min.}}{2} - \left\{ 1\cdot8 + \frac{a(b-c)}{30} \right\} = \text{the}$$

mean temperature of the day.

Where the mean temperature of the day is only required to the nearest half-degree, the monthly correction, without alteration, may be applied to every day's mean of the maximum and minimum; this will give the mean temperature of the day sufficiently near to serve the purpose of tracing connexion between change of temperature and the progress of disease; in the consideration of which it enters as an important element.

Now, on the supposition that these quantities are the same for the whole of Great Britain, we have two independent methods of arriving at the mean temperature of the month, which may serve mutually to correct each other; that they are of this extensive application will not admit of proof until we have more registers of temperature taken frequently in the twenty-four hours than exist at present; if, however, we find that the mean temperatures derived from observations at set hours in the day agree in the main with those resulting from the means of the maximum and minimum observations at any place, we may fairly conclude that these corrections apply as well to that locality as to Greenwich.

54. **Example.**—To test the application of these corrections, we will take the maxima and minima observed at Southampton in January 1849, the month before quoted; I find the mean of the maximum observations 45·3, of the minimum 37·4.

The mean temperature deduced from these, with Glaisher's corrections applied, will be

$$\frac{45{\cdot}3+37{\cdot}4}{2}-0{\cdot}2=41{\cdot}15,$$

which differs from the former result obtained in § 51, by 0°·4.

Now the mean temperature derived from the maxima and minima results from two observations daily; in determining therefore the mean temperature from a combination of all our observations, we give this determination double weight; and thus the "adopted mean temperature" would be that derived from the following formula, in which are combined, with the maxima and minima, the observations taken at 9 A.M., 3 P.M., and 9 P.M.—

$$\frac{41{\cdot}67+41{\cdot}66+41{\cdot}69+2(41{\cdot}15)}{5}=41^{\circ}{\cdot}46 \text{ M. T.}$$

The results of the two methods of determining the monthly mean temperature will be seen in the accompanying Table; they are from the author's

	1848.		1849.		1850.	
	M. T. from three daily obs.	M. T. from Max. and Min.	M. T. from three daily obs.	M. T. from Max. and Min.	M. T. from three daily obs.	M. T. from Max. and Min.
January........	......	......	40·1	39·8	34·9	33·8
February	43·3	43·3	43·6	42·7	44·6	43·8
March...........	43·6	46·6	43·2	43·1	40·4	38·7
April	47·7	49·1	44·4	44·6	48·4	48·6
May	58·7	58·1	57·1	58·8	51·6	52·5
June	56·7	59·4	59·5	60·7	59·5	59·2
July	61·7	62·2	60·8	59·9	61·6	62·1
August	58·5	59·6	61·	60·5	59·5	60·2
September	56·9	57·2	56·3	57·	56·3	55·5
October	51·9	51·7	51·6	51·	46·6	46·2
November	42·8	42·6	45·5	44·8	47·1	46·7
December	44·1	44·	39·9	38·7	42·8	42·0
Means for the year ...	51·4	52·1	50·2	50·1	49·4	49·1

Mean temperature from three observations daily during the three years .... 50°·3

Mean temperature from the maxima and minima......... 50·4

three years' observations taken three times daily, viz. 9 A.M., 3 P.M., and 9 P.M. The comparison will show a sufficient accordance to give a confidence in the use of the Table of Corrections, § 48, which was originally calculated for the Royal Observatory at Greenwich.

Feeling a confidence in the use of the Greenwich corrections for diurnal range from these observations, I gave up the two later observations in the day after the year 1850, and confined myself to 9 A.M.; the results for the following three years are here given:—

	1851.		1852.		1853.	
	M. T. from 9 A.M. observations.	M. T. from Max. and Min.	M. T. from 9 A.M. observations.	M. T. from Max. and Min.	M. T. from 9 A.M. observations.	M. T. from Max. and Min.
January	44·3	44·1	42·	42·5	43·4	43·5
February	40·6	40·8	41·2	42·3	35·	35·8
March	44·3	43·3	42·1	42·1	38·	38·6
April	46·4	45·	46·8	46·	47·1	47·6
May	51·6	50·3	51·1	52·6	52·6	52·8
June	58·3	56·8	56·4	56·9	57·	57·7
July	61·5	61·3*	65·9	65·7	59·9	59·1
August	62·2	63·	61·2	61·4	61·2	59·8
September	57·5	57·6	56·8	56·9	55·9	55·9
October	53·8	53·4	48·1	78·	52·5	50·4
November	38·5	39·7	49·9	50·2	42·6	44·3
December	41·4	41·9	48·2	47·8	36·6	37·7
Means for the year ...	50·	49·7	50·9	51·	48·5	48·6

Mean temperature from the 9 A.M. daily observations during the three years........................ 49°·8

Mean temperature from the maxima and minima ...... 49·8

Mean temperature of Southampton from the six years' observations .... 50·0

* Interpolated.

55. Explanations.—It may be useful, in guiding those who are about to commence a series of observations, to make some remarks on the above observations. The instruments with which the observations were taken, were good of their kind, but not expensive; the thermometer used at the stated hours was compared with a standard by immersion in warm water, frequent readings being taken at intervals of ten minutes or a quarter of an hour, as the water gradually cooled; the sum of the readings of each thermometer divided by the number of readings gave a mean temperature, which, if the readings of both thermometers had been the same throughout, would have been the same quantity; the mean reading of my thermometer was, however, 0°·4 higher than the standard of comparison, hence to all my readings a correction of —0°·4 has been applied to reduce them to true readings. The maximum and minimum thermometers were, in like manner, compared with the other corrected, and the index correction applied throughout.

The thermometers are only divided to degrees; hence, in reading off, if the scale is cut by the mercury between two degrees, we estimate how many tenths of a degree it is above the next lowest, and enter the reading accordingly.

It will be remarked, and the remark will apply

to all other observations, that though the differences, in some individual months, between the mean temperature derived from observations taken at certain hours and corrected for diurnal range by Glaisher's Tables, and that deduced from the mean of the maxima + the mean of the minima ÷ 2 — the quantities given in § 53, is considerable, these all vanish in a series of no very great extent. Rejecting 1848, which is an incomplete year, the difference between mean temperatures of the year by the two methods will be found to be:—

50·2—50·1 or +0·1 difference for 1849
49·4—49·1 or +0·3 difference for 1850
50· —49·7 or +0·3 difference for 1851
50·9—51·0 or —0·1 difference for 1852
48·5—48·6 or —0·1 difference for 1853

56. Projection of six years' observations.—An excellent method of recording phenomena, so as to present them to the eye at a glance, is that of projecting them in curves, taking the time for abscissæ and the variables for ordinates; thus the monthly temperatures recorded in the two preceding Tables may be projected in the form shown in Plate III., in which the monthly irregularities are rendered visible. In addition to these, the same elements for Greenwich are projected on the same scale, for the purpose of ready comparison, as also the normal curve of monthly temperature at the Royal Observatory, Greenwich, derived from seventy-nine years of observation; the data being the sum of the temperatures for each month of the same name, divided by the number of years through which the observations extended.

From all the observations combined, the mean temperature of each month at the Royal Observatory, Greenwich, is—

of January ...	35·7	of July	61·3
of February ...	38·2	of August	60·5
of March	40·9	of September ...	56·3
of April	45·7	of October	49·3
of May	52·6	of November ...	42·4
of June.........	58·0	of December ...	38·8

The mean of all the monthly results or mean temperature for the year is 48°·3.

Height of the Observatory above the mean level of the sea, 160 feet.

The height of my observatory above the sea-level was only 60 feet; the difference between this and the height of the Royal Observatory, is sufficient to account for the slight excess of cold in that position during the winter months; the proximity of Southampton to the sea is indicated in the extremes of temperature being somewhat less than at Greenwich.

57. Care necessary in observation.—It may be

stated as an indisputable fact, that unless great care be taken in choosing a position for the thermometers, the whole of the observations will be vitiated; many sets of records have proved worthless from want of attention to this important matter, and much valuable labour has been thrown away, not only from this circumstance, but also from the use of inferior instruments. It is true, that, by a careful comparison with a standard throughout the whole extent of the scale, quantities might be obtained which, applied to the observations made with a bad instrument, may bring them near the truth; but if the observations have been taken in a place where reflected heat has been recorded, or where the sun's rays may have had an influence, no good results can possibly be obtained.

58. **Position of instruments.**—The corps of observers in England who combine their observations, as will be explained more fully in Part III., suspend their thermometers in positions and under circumstances as much as possible similar: the bulb of the dry thermometer is supposed to be 4 feet from the ground, and the temperature shown by it is due to the stratum of air at that distance from the soil; it is sheltered in every way, not only from the direct rays of the sun, but it is also so far removed from walls and buildings as to be

out of the way of the reflected heat, or that derived from radiation; it must be protected from rain, but the air must be allowed an uninterrupted passage around it in every direction. The author has endeavoured to ensure all these advantages by a stand of the following construction.

59. **The Author's thermometer stand.**—In it are suspended the dry-bulb and wet-bulb thermometers, and also the thermometers which register the greatest and least degrees of heat. The stand is doubly boarded on the east, west, and south sides, the instruments facing the north without any screen; hence when the sun shines on the outer covering, a stratum of air intervening, the inner case does not get heated by its rays; holes are bored in various parts of both the inner and outer case, with especial care that they be not opposite to each other, lest the sun's rays should find admission to the thermometers; the air thus circulating freely is supposed to be of an equal temperature with that in the shade in any position, however sheltered from the sun: a pent-house, or sloping roof, also double, throws off the rain, and thus the thermometers are kept dry. The upright support of the stand is firmly fixed in the ground, and braced to prevent agitation; the whole should be at least 12 or 15 feet from lateral walls, and 40 or 50 from walls directly north,

which would reflect the sun's rays and influence the instruments. The thermometers face the north and are about 4 feet from the ground. Without some such stand, it must be impressively stated, any thermometric observations will be useless, as the instruments will not be protected from local or celestial interferences which would entirely vitiate the most careful records; nor would the register be comparable with observations taken in different parts of the country.

60. Greenwich thermometer stand.—The thermometer stand after the form of that at Greenwich, and, till lately, at Kew, differs from this, as may be seen in Plate IV. fig. 1.

A is a wooden painted frame, 2 ft. 6 in. square, a^1 a base board attached to the under part of A, a^2 the underside of a strong piece attached to the upper part of A, a^3 an interior portion of a penthouse composed of very thin boards attached to a^1 and a^2, a^4 a^5 strips of wood to which the thermometers are attached.

B is a strong spar or post firmly secured to the ground at Greenwich, to a stone balustrade at Kew; in its upper end is fixed a cylindrical pin which freely enters a socket in the central part of a^2, and allows A, with all its adjuncts, to be revolved on the axis of B.

Now the great objection to this kind of stand

for general use is that it requires, during the summer time when the sun's azimuth is north of east or west, to be turned bodily, so that the roof be directed to his rays which would otherwise impinge laterally on the thermometers. The problem to be solved in the construction of a stand is that the instruments be at all times protected from the direct influence of the sun and the rain, at the same time that they are freely exposed to unimpeded circulation of air. The stand erected at Kew by Mr. Welsh is perhaps as near an approach to the solution as has yet been attained, having the recommendation of requiring no adjustment after it has once been fixed in the ground, being open to the movement of the air all round, and shielded from the approach of the sun's rays whatever may be his azimuth.

The thermometers are in a cubical case, the sides being about 2 feet, open at the bottom; it is composed of louvre-boards, like Venetian shutters. Another case of the same construction, but of twice the dimensions every way, encloses the smaller. The cases reach to within 4 feet of the ground, so that there is nothing to obstruct the free passage of the air horizontally, which moreover finds its way readily through the open woodwork, while the sun can only reach the external case and not that in which the instruments are suspended.

61. Mean yearly temperature at Greenwich.—In the 'Philosophical Transactions,' Part II., 1849, and Part II., 1850, will be found two most valuable papers by Mr. Glaisher, in which he has reduced observations made at the Royal Society's Apartments Somerset House, at Epping, and at Lyndon in Rutlandshire, to an agreement with those made at the Royal Observatory, Greenwich; this series extends from the year 1771 to 1849, and from it that gentleman has determined the mean temperature of every year, at Greenwich, through a period of seventy-nine years. Now as Greenwich is ever likely to be a point of reference in all observations on climate, and as the difference in temperature between it and any other place in England is readily ascertained, we consider this series to be one of great value; it is probably the best determination of this element ever completed, and it will assist in arriving at other valuable results; the number of observations treated of exceeds 200,000, spread nearly equally over the seventy-nine years. Mr. Hugh Gordon, of the Ordnance Map Office, Southampton, having been employed in discussing these observations, determined to equate the several yearly temperatures by an elliptic curve, starting from one lowest point to the next. He has kindly supplied the author with the projection, engraved in Plate V., whose elliptically equated curve shows, in a very striking manner, the cycle of the greatly variable mean temperature at the Royal Observatory. The numbers below the lines show the mean temperature of each year as deduced from observation, the equated values may be ascertained by inspection; it will be found that the summation of the equated temperatures in each cycle equals the summation of the recorded mean temperatures within the same period.

An inspection of the form assumed by this curve will exhibit certain remarkable phenomena; commencing with 1771, the years gradually became warmer till 1779, when the temperature in like manner declined, and a batch of cold years occurs, of which 1784 was the coldest. The heat then increased, but not in so great a degree, till 1794, when the extreme cold of that cycle, not so severe as before, was reached gradually in five years from that time. In periods varying from nine to fifteen years throughout the whole series, we find the cycle of hot and cold years repeated; and nothing can more distinctly present the recurrence of these phenomena to the mind than the graphic method here adopted. In this projection, it is indicated to the eye in the sequence of the repeated elliptic curves.

62. Comparison with that of other places.—By examining closely and comparing the meteorological reports from various places in England, Mr. Glaisher has endeavoured to arrive at some empirical formula, by which, the mean yearly temperature at Greenwich being known, that of any other place may be obtained; he takes into account the law of decrease of heat with increasing elevation, and the difference of latitude, and the results, in nearly every case, have very closely approximated to those of observation, which in some instances they have tended to correct, since a want of accordance between the two, has been found, more than once, to be due to errors either in the instruments employed or in the observations made with them.

Let the mean temperature of the air at the level of the sea at Greenwich, latitude 51°·5, be denoted by T; on the supposition that this value, at a uniform level, becomes less by 0°·9 for an increase of latitude of one degree; then—

The mean temperature of any place in England for that year may be calculated approximately from the following formula:—

$$T + (51^\circ{\cdot}5 - \text{lat.}) \times 0{\cdot}9 - {\cdot}00345 \times \text{height of the place in feet above the level of the sea.}$$

By substituting the mean temperature of Greenwich for any year, the formula is of perpetual application; thus, from my observations I find the mean temperature of Southampton for 1853 to be

48°·5, the latitude being 50°·9, and the height above the sea 60 feet; the mean temperature of Greenwich for the same year reduced to the sea-level=48·2; by substitution,

$$48{\cdot}2+(51^\circ{\cdot}5-50^\circ{\cdot}9)0{\cdot}9-{\cdot}00345\times 60=48^\circ{\cdot}5,$$

a result exactly coincident with that deduced from the observations.

63. Mean temperature fails in describing climate.—Mean temperature, whether of the year or month, though an important element in investigation, conveys but a very remote idea of the climate of any locality. The average degree of heat will remain uninfluenced by sudden temporary rises and corresponding depressions, so that the result for a place subject to extreme variations may be the same as for one whose temperature is equable. The differences in this element in various localities may arise from four different causes:—*a*. Distance from the equator, which is the cause of the diurnal arc of the sun varying with the season. *β*. Height above the sea-level, which will cause, at moderate elevations, a decrease in temperature of about 1° for every 300 feet. *γ*. The geological character of the neighbourhood. *δ*. Situation as regards the sea-coast. Of these causes, the last produces the greatest irregularities, and such as are most difficult to reduce to system.

64. Mean daily temperature at Greenwich.—The average daily temperature for Greenwich, which probably varies according to a law that is the same for a great part of England, has been determined from all the thermometrical observations taken at the Royal Observatory during 43 years, ending with 1856. Tables of the adopted mean temperature of every day in the year and of the data from which this was computed, are to be found in a paper by Mr. Glaisher in the Report of the Council of the British Meteorological Society for 1857. The Tables show that the temperature does not regularly increase from its lowest in the middle of January up to its greatest height in July, but that there are variations of some days' duration in the months of February, March, and May: its decline goes on regularly from the end of July till the end of November, when an increase takes place which just lasts into December, when it again falls, though it is somewhat irregular all through this month.

Another of the Tables gives factors for each day of the year, which, when multiplied into the mean temperature of each month, will give the mean temperature of any particular day of the month. It is by these numbers that the results of the Greenwich observations are made applicable to other places; for it is likely that in many localities the average temperature of a day bears the same proportion to the mean monthly temperature as it does at Greenwich. Another Table shows the distribution of heat over the year in the following manner: the average amount of heat is represented by 1 (which number occurs at the 29th of April and the 21st of October), and for each day a decimal shows the ratio the number of degrees representing its mean temperature bears to that of the mean yearly temperature; the lowest fraction is (for January 15 and 16) ·726, and the highest (for the days from July 28 to August 1) 1·279. It must not, however, be supposed that this tells us the absolute proportion of heat received on each day, for it is clear that the numbers would be different if Reaumur's or the Centigrade scale were used, from their having a different zero, and therefore they depend on an arrangement that is artificial. In order to show the true proportion of the differences of temperature on the different days, the Editor has constructed the following Table, taking for data the adopted mean temperature of every day of the year, given in one of those mentioned above. Here the lowest daily temperature is the zero, the highest is unity, and the decimals tell what part of the whole increase of daily temperature that takes place during the year has occurred at the date opposite each number.

65. Table showing how the heat increases through the year.

Days of the Month.	Jan.	Feb.	March.	April.	May.	June.
	o	o	o	o	o	o
1	·037	·063	·170	·300	·537	·774
2	·033	·055	·166	·318	·555	·781
3	·033	·066	·162	·333	·570	·788
4	·029	·081	·162	·344	·585	·800
5	·022	·107	·166	·355	·596	·800
6	·018	·130	·170	·366	·613	·803
7	·011	·137	·170	·370	·607	·807
8	·001	·137	·170	·370	·603	·811
9	·000	·130	·174	·366	·596	·814
10	·007	·118	·177	·355	·588	·822
11	·003	·111	·185	·351	·585	·833
12	·003	·103	·188	·348	·585	·844
13	·003	·100	·203	·355	·588	·851
14	·003	·092	·214	·370	·603	·862
15	·000	·096	·222	·377	·618	·870
16	·000	·096	·229	·388	·633	·881
17	·014	·100	·233	·396	·640	·885
18	·029	·103	·233	·403	·651	·888
19	·037	·107	·233	·407	·659	·892
20	·048	·111	·237	·414	·670	·900
21	·059	·111	·237	·425	·677	·903
22	·066	·114	·237	·444	·688	·907
23	·074	·126	·240	·448	·696	·914
24	·085	·133	·244	·448	·700	·922
25	·092	·148	·248	·448	·707	·929
26	·103	·155	·251	·448	·711	·937
27	·103	·159	·255	·462	·718	·944
28	·096	·166	·262	·481	·729	·955
29	·088	..	·270	·503	·737	·962
30	·077	..	·277	·518	·748	·962
31	·066	..	·292	..	·762	..
Means	·040	·112	·215	·397	·643	·868

Table continued.

Days of the Month.	July.	August.	Sept.	October.	Nov.	Dec.
	°	°	°	°	°	°
1	·962	1·000	·862	·666	·403	·229
2	·959	·992	·855	·662	·396	·233
3	·959	·988	·851	·651	·392	·229
4	·962	·985	·848	·648	·385	·222
5	·966	·981	·840	·640	·377	·218
6	·970	·981	·833	·629	·370	·207
7	·974	·981	·825	·622	·355	·192
8	·970	·981	·822	·614	·344	·188
9	·962	·977	·818	·607	·333	·185
10	·962	·974	·814	·600	·325	·181
11	·966	·970	·811	·588	·314	·174
12	·970	·966	·807	·574	·311	·166
13	·974	·962	·803	·555	·296	·151
14	·970	·962	·796	·540	·288	·155
15	·970	·959	·785	·529	·277	·166
16	·970	·959	·777	·522	·262	·177
17	·970	·955	·770	·514	·255	·170
18	·970	·944	·759	·507	·248	·162
19	·970	·940	·755	·503	·248	·148
20	·966	·937	·744	·503	·248	·129
21	·962	·929	·740	·496	·240	·107
22	·962	·922	·737	·485	·229	·088
23	·962	·918	·733	·474	·218	·070
24	·966	·918	·722	·459	·203	·055
25	·974	·914	·718	·444	·196	·037
26	·985	·903	·711	·440	·200	·033
27	·992	·896	·703	·433	·207	·037
28	1·000	·892	·696	·425	·222	·055
29	1·000	·885	·685	·418	·225	·066
30	1·000	·874	·677	·411	·225	·070
31	1·000	·870	..	·407	..	·077
Means	·972	·945	·776	·534	·286	·141

66. Dove's Isothermal Charts.—Professor Dove has given great attention to the subject of mean temperatures, and his valuable Tables, published in the Report of the British Association for 1847, with his Isothermal Charts (described in that for 1851) for every month of the year, will well repay consultation. In the formation of these, 900 places have been selected at which continuous observations have been made; these have all been reduced to the level of the sea, and the temperatures of other places have been determined by empirical formulæ; lines are then drawn across a map of the world through those places whose monthly mean temperature is the same. From mere inspection of these we learn some very important particulars, the discussion of which in full would be foreign to the purpose of this work, though we should not be justified in passing over altogether certain conclusions at which Professor Dove has arrived.

67. Deductions from them.—*a.* He has shown that the mean temperature of the western hemisphere—supposing the meridian of Faro to be the separation—exceeds that of the eastern for every degree of latitude except 70°, but that the difference is less as the places approach the equator, which is accounted for by the preponderating mass of water in the west. There is no point in the western hemisphere which approaches the extreme cold of Yakutsk, whose January temperature is —40°.

β. The winters of the northern hemisphere are colder than those of the southern; on the contrary, the summers are warmer. Taking the mean heat of all the places on the globe in January and July, he calculates that of the entire earth to be—

January..	In the northern hemisphere	7°·5 R.	=	48°·9 Fahr.
	In the southern	12·2 R.	=	59·5 F.
	Mean of the whole earth ...	9·9 R.	=	54·3 F.
July	In the northern hemisphere	17·3 R.	=	71 F.
	In the southern	9·6 R.	=	53·3 F.
	Mean of the whole earth ...	13·5 R.	=	62·4 F.

The temperature of the earth, therefore, increases 8° from January to July.

The preponderance of water in the southern hemisphere will account for the less variable degree of heat to which it is subject. The increase of the temperature of the globe between the winter and summer of the northern hemisphere, is the result of the unequal distribution of land and sea in the two hemispheres, and the difference of the effect produced by the rays of the sun, according as they impinge on a solid or a liquid surface.

γ. The difference between the hottest and

coldest months of the year is strikingly exhibited in Dove's Maps; the range of temperature between the months of January and July, according as it is small or great, produces those differences in atmospheric phenomena, to which may be given the names of "continental" and "insular" climates. As extreme examples we may cite Commewine in Guiana, where the temperature in July differs from that in January only 2°·2; and Yakutsk in Siberia, where the variation in temperature of these months is 114°·4.

68. Continental and insular climate compared: Example.—To illustrate the important distinction of continental and insular climates, I may select an example from a series of observations, with which, for some years, I have been favoured by W. Palk, Esq., M.D., of Union, Franklin Co., Missouri, United States, Lat. 38° 36′ N., Long. 91° 10′ W., in the very centre of North America. They have been taken with great care three times daily, viz. at 9 A.M., 3 P.M., and 9 P.M.; the thermometer stands facing the north under a portico, in such a position as will render safe the comparison of his observations with my own.

I find the approximate mean temperature of Union for January 1853 to be 36°·2, which differs but slightly from the normal mean temperature of that month at Greenwich, and agrees remarkably with the January temperature given to its position in North America on Dove's chart, which is so far a proof of the accuracy of his empirical formula.

When, however, we examine the daily observations, we shall find that the range of temperature is so extraordinary that we experience no variations at all corresponding with them; thus on the 7th, 9th and 10th of the month, the thermometer at 3 P.M. stood at 70°; at 9 A.M. of the 3rd it showed 10°; the monthly range was then 60°; the corresponding monthly range at Southampton, which exhibits every symptom of an insular climate, was 55° — 32° = 23°. The following extracts from the register will show the variable character of the continental climate:—

		9 A.M.	3 P.M.	9 P.M.
January	3	10°	20°	18°
"	7	52	70	54
"	10	58	70	48
"	26	12	26	24
"	27	22	50	26

The 7th of January, at Union, must have been like a pleasant summer's day, while on the 3rd the cold was greater almost than we ever experience; on the 27th, the range of the thermometer in six hours was 28°; and in the following six, the air cooled down 24°.

The mean heat of the month of July was about 76°; the range of mean temperature between that and January was 40°; the same element at Southampton was 16°. These were the observations on some of the hottest days in the year 1853:—

	9 A.M.	3 P.M.	9 P.M.
	°	°	°
July 1	82	100	80
„ 4	84	96	76
„ 7	76	96	74
„ 8	86	102	80
„ 14	86	100	86

We see, then, that a continental climate deals in extremes, especially during the winter months; and it is clear that we learn little of its peculiar features from the mean value of either yearly or monthly temperatures; we must descend further into particulars, and mark the sudden transitions from heat to cold and the reverse before we can judge of the peculiarities due to any particular locality.

These extreme changes of temperature are very trying to the constitution, and diseases of a certain character are prevalent during extraordinary transitions. "An old practitioner," adds Dr. Palk, "can always tell what diseases will follow a sudden change of weather, whether to hot or cold."

It is interesting to compare climates differing greatly from each other; we will therefore inspect the register of temperatures kept at Alten in Norway, latitude 70° north.

	1846.	1847.	1848.
	°	°	°
Maximum ...	83·3	84·7	86·9
Minimum ...	−14·8	− 3·1	− 20·2
Range	98·1	87·8	107·1

This place, although situated on the sea-coast, has such an extreme range of temperature that its climate may be designated as "continental," a word which, as well as "insular," is frequently applied to climate by meteorologists, without reference to the position of a place as regards the sea-board. In this case it will be seen, by a reference to the map of Europe, that Alten is far removed from the influence of the Gulf-stream, which modifies the climate of all those places which it reaches in its course.

69. Temperature affected by the neighbourhood of water: Example.—The fact of the neighbourhood of water tending to equalize temperature is shown in localities where even a very limited extent of surface, as that of a river, is able to influence the superjacent air. In a comparison before alluded to, undertaken in the course of the investigations about the degree of temperature due to Epping, Lyndon, and Somerset House, it was found, after reduction to the sea-level, that the thermometer at Somerset House indicated a higher winter temperature than at Epping by about 1°·4, and a lower summer temperature by about 0°·9; and it was concluded that this variation was due to the vicinity of the river Thames. For many years the temperature of the water has been recorded by a maximum and a minimum thermometer 2 feet below the surface of the water of the Thames, suspended from the sides of the Dreadnought Hospital Ship, in a perforated trunk. The results of these observations, undertaken by Captain Saunders, R.N., are distinctly shown in the following account by Mr. Glaisher:—"By comparing the mean monthly temperature of the water of the Thames for the four years (1846, 1847, 1848, 1849) with the means of the readings of the maximum and minimum thermometers in air at the Royal Observatory, for the same months, we find that the mean lowest readings of the water were higher in the twelve months respectively by 3°·9; 4°·6; 7°·7; 12°·3; 11°·5; 16°·7; 12°·4; 10°·4; 10°·7; 6°·8; 6°·3; and 4°·9, than the mean of the lowest readings of the air; and it was lower than the mean maximum readings of the air by 3°·2; 4°·6; 5°·3; 9°·2; 6°; 7°·5; 8°·1; 6°·7; 7°·4; 6°; 4°·7; and 3°, in the respective months from January. These numbers are very large, and will fully account for the little higher temperature possessed by places in the vicinity of the river; and these differences of temperature are probably the fruitful source of the London fogs."

We can very directly trace the effect of the presence of water upon temperature by looking at the readings of a thermometer in the air at the Dreadnought itself, and comparing them with those taken under similar circumstances at Greenwich; this is done in the following extract from the same paper of Mr. Glaisher's:—"The temperature of the air 32 ft. above the water (from observations at 6 A.M. and 6 P.M. in 1847 and 1848, and at 9 A.M. and 9 P.M. in 1849) exceeds that at the Observatory (height 160 feet) at 6 A.M., by 1°·6; 1°; 0°·8; 0°·3; 0°·6; 0°·7; 0°·9; 0°·8; 0°·2; 0°; and 0°·8, in the twelve months respectively; and at 6 P.M. by 1°·2; 0°·8; 1°·0; 0°·8; 0°·7; 0°·8; 0°·6; 0°·8; 1°·3; 1°; 1°·7; 0°·9, in the twelve months respectively; at 9 A.M. it was in excess in January by 1°·3; February by 1°·5; March by 0°·6; April by 0°·4; May by 2°·2; June by 0°·4; and in October by 0°·5; it was of a lower temperature in July by 0°·7; August by 0°·5; and in September by 0°·1; at 9 P.M. it was always of a higher temperature: the excesses were 0°·1; 0°·3; 0°·7; 0°·3; 1°·9; 2°·9; 1°·5; 3°·2; 1°·2; and 1°·3 respectively."

"From these numbers, it seems that during the night hours, at all seasons of the year, the temperature of the air at the Dreadnought Hospital Ship is higher than at the Observatory, and that it is below only during the mid-day hours."

"At times of extreme temperature, the effect of the water upon the temperature of the air is very great. On Feb. 12, 1847, the temperature of the air, at my house situated one mile and a half from the river, was 6°; the lowest reading 32 feet above the water of the Thames was 16°; the temperature of the water was 33; its heating effect upon the air in its immediate vicinity amounted to 10°; at the Observatory the reading was 10°·5, and the heat of the water of the Thames seems to have influenced the temperature of the air at the Observatory to the amount of 4°."

70. Decrease of temperature as height from the sea-level increases.—The following Table, derived from the Report of the Balloon ascent mentioned in § 5, will show the decrease of heat as the aëronauts rose in the air, on Aug. 17, 1852, between 4 and 5 P.M.:—

Heights above the level of the sea.	Temperatures.	Heights above the level of the sea.	Temperatures.
ft.	°	ft.	°
120	71·2	5,880	57·8
2,440	62·8	6,800	54·0
3,460	59·2	7,530	51·4
4,110	58·1	8,550	49·0

Heights above the level of the sea.	Temperatures.	Heights above the level of the sea.	Temperatures.
ft.	°	ft.	°
9,470	44·4	15,510	24·4
10,680	40·4	16,600	20·6
11,620	37·2	17,440	19·6
12,250	34·9	18,490	15·0
13,480	30·8	19,320	10·5
14,550	27·0		

Mr. Welsh, who discussed the whole series of observations, comes to the conclusion, "that the hypothesis of a regular progression," in the decrease of temperature, "at all heights, can scarcely be maintained." The results of the observations of decrease of temperature were discordant in the four balloon ascents, and we must wait for further observations in elevated regions, before we can arrive at satisfactory conclusions as to the thermometric state of the upper strata of the air.

71. Effect of the sun's heat below the surface of the soil.—An important difference of temperature between places otherwise similarly situated will arise from their geological formations; those on arid dry soils will have a greater range of temperature than those situated on a clayey soil, or any other which prevents the drainage or carrying away of the water received in the form of rain. To assist in arriving at distinct conclusions on this head, and to judge of the depth to which the sun's rays are able to penetrate, and of the radiating power of earthy materials, thermometers, constructed for the purpose, should be sunk in the ground and registered at stated hours. During the year 1850, at Southampton, the author kept a daily record of the readings of a thermometer near the surface of the soil, which was garden mould, and of another sunk one foot below. They were read simultaneously at 3 P.M.

Readings of a thermometer near the surface and another one foot below, at Southampton.

1850.	Therm. at the surface.	Therm. one foot below the surface.	1850.	Therm. at the surface.	Therm. one foot below the surface.
January ...	35·8	35·4	July	69·5	61·3
February...	45·	40·8	August ...	66·1	59·1
March......	45·8	39·2	September	63·7	54·5
April	54·4	47·	October ...	51·1	47·1
May	58·7	47·8	November .	47·4	46·
June	68·8	61·5	December .	41·1	40·8

M. Quetelet has discussed this subject very fully in the "Memoir" referred to in § 48; founding his remarks on observations taken between 1834 and 1842; to illustrate it we will, however, look nearer home.

72. Observations at Greenwich.—Resorting to the Greenwich observations for 1847, we find some very striking results with regard to the range of heat at different distances below the earth's surface, which are best exhibited in a tabular form.

Table I.—Mean monthly reading of a thermometer whose bulb is placed on a level with the scales of the deep sunk thermometers, from daily observation in 1845.

	Jan.	Feb.	Mar.	April.	May.	June.
Mean monthly reading	36·4	36·4	42·	46·	37·4	59·2
Mean daily range	6·9	8·4	14·3	13·6	17·1	16·9
	July.	**Aug.**	**Sept.**	**Oct.**	**Nov.**	**Dec.**
Mean monthly reading	66·9	63·4	55·	53·2	47·3	42·9
Mean daily range	19·9	17·1	15·3	11·4	9·4	7·4

Table II.—Mean monthly reading of a thermometer whose bulb is sunk 1 inch below the soil.

	Jan.	Feb.	Mar.	April.	May.	June.
Mean monthly reading	37·9	38·	42·3	46·4	56·7	59·8
Mean daily range	4·1	4·4	7·	6·9	9·4	8·1
	July.	**Aug.**	**Sept.**	**Oct.**	**Nov.**	**Dec.**
Mean monthly reading	66·8	63·9	56·	53·7	48·4	44·
Mean daily range	10·1	9·7	7·4	5·9	5·	4·5

Table III.—Mean monthly reading of a thermometer whose bulb is sunk 3·2 ft. (=3 French feet.)

	Jan.	Feb.	Mar.	April.	May.	June.
Mean monthly reading	39·3	39·6	41·1	44·4	51·2	57·2
Mean daily range	0·4	0·3	0·3	0·3	0·4	0·4

	July.	Aug.	Sept.	Oct.	Nov.	Dec.
Mean monthly reading	61·5	62·3	57·9	54·7	50·6	46·4
Mean daily range	0·5	0·4	0·3	0·3	0·3	0·3

Table IV.—Mean monthly reading of a thermometer whose bulb is sunk 6·4 ft. (=6 French feet.)

	Jan.	Feb.	Mar.	April.	May.	June.
Mean monthly reading	44·	42·8	43·3	45·1	48·6	53·9
Mean daily range	0·07	0·12	0·11	0·13	0·22	0·21

	July.	Aug.	Sept.	Oct.	Nov.	Dec.
Mean monthly reading	57·	58·9	57·8	55·3	52·8	49·5
Mean daily range	0·36	0·17	0·14	0·15	0·14	0·11

Table V.—Mean monthly reading of a thermometer whose bulb is sunk 12·8 ft. (=12 French feet.)

	Jan.	Feb.	Mar.	April.	May.	June.
Mean monthly reading	49·	46·7	45·8	45·8	46·8	49·5
Mean daily range	0·12	0·09	0·09	0·10	0·11	0·14

	July.	Aug.	Sept.	Oct.	Nov.	Dec.
Mean monthly reading	51·9	54·2	55·11	54·6	53·6	51·8
Mean daily range	0·15	0·16	0·13	0·10	0·08	0·09

Table VI.—Mean monthly reading of a thermometer whose bulb is sunk 25·6 ft. (=24 French feet) below the surface of the soil.

	Jan.	Feb.	Mar.	April.	May.	June.
Mean monthly reading	52·	51·2	50·2	49·3	48·8	49·7
Mean daily range	0·05	0·04	0·05	0·05	0·05	0·04

	July.	Aug.	Sept.	Oct.	Nov.	Dec.
Mean monthly reading	49·0	49·8	50·6	51·3	51·7	51·8
Mean daily range	0·06	0·07	00·6	0·06	0·05	0·05

We see from these Tables that the yearly extremes occur in a later month the deeper the thermometer is placed, until, in the last instance when the thermometer bulb was 25½ feet below the surface, the highest reading occurred in January and the lowest in July.

Also we see distinctly that the deeper we descend below the earth's surface, the more equable is the temperature; at every place we should at last arrive at a depth where the action of the sun's rays would not be felt at all, but a thermometer would show the same degree of heat at all times and seasons.

73. Stratum of invariable temperature.—The "stratum of invariable temperature" will be at different depths at different places.

From M. Quetelet we extract the following

table of depths at which the annual variation of the thermometer is reduced to 0°·01 Centigrade: the feet are French, 1 foot French being equal to 1·0658 English.

	ft.		ft.	
Zurich.........	83·7	Leith.........	54·7	
Strasburg ...	81·6	Edinburgh...	55·5	Trap.
Heidelberg ...	83·3	ib.	66·2	Sand.
Schwetzingen	89·8	ib.	96·6	Sandstone.
Bonn	72·6	Upsal	62·6	
Paris	69·4	Trevandrum	53·6	

Below the "stratum of invariable temperature" the heat of the earth would seem to increase in different proportions, for the most part at the rate of 1° for 72 feet of depth; but as this subject does not bear directly upon meteorology, we shall not carry the discussion of it further.

74. **Radiating thermometers.**—A very important element, registered at Greenwich, is the amount of "solar radiation," the measurer of which is a self-registering mercurial thermometer with a blackened bulb; it is placed on the ground in an open box, the sides of which are sufficiently high to prevent lateral wind from striking the bulb. The degrees are marked on the tube itself, to prevent accumulation of heat, reflexion, and radiation from a scale of wood or metal

An arrangement likely to give yet more accurate results is to enclose the bulb of the thermometer in a glass globe, from which the air is exhausted; this will prevent the lowering of the temperature by currents of air, which, there is reason to believe, have brought the radiation thermometer down below its proper reading.

For the "sky-radiation" or terrestrial radiation, that is, radiation from the earth upwards, a self-registering minimum thermometer is used; its bulb is placed in the focus of a parabolic metallic reflector, which is turned upwards towards the clear aspect of the sky and screened from currents. Even in the day-time a thermometer so placed and turned towards the clear sky, but away from the rays of the sun, will fall several degrees below the temperature of the surrounding air.

It would be well, also, to observe a minimum thermometer placed near it and exposed to the sky too but without the mirror.

75. **Actinometer.**—An instrument, termed the Actinometer, devised by Sir John Herschel, has been extensively used to ascertain the absolute heating effect of the solar rays, in which *time* is considered as one of the elements of observation.

The actinometer consists of a large cylindrical thermometer-bulb, with a scale considerably lengthened, so that minute changes in temperature may be easily appreciated. The bulb, which is of transparent glass, is filled with a deep blue liquid, which is of course expanded when the rays of the sun fall direct on the bulb. To take an observation with the actinometer, it is exposed to the sunshine for one minute and read off; it is then withdrawn to the shade for a minute and its indication recorded; it is then again placed in the sun and this alternation continued for any time, care being taken to begin and end the series with a sun observation; the mean of the readings in the shade, subtracted from the mean of those in the sun, gives the actual amount of dilatation of the liquid produced by the sun's rays in one minute of time. The mode of registration is extensively explained in the Report of the Committee of the Royal Society on Physics and Meteorology, to which the reader is referred for further information.

Very interesting observations may be made by two observers simultaneously with actinometers previously compared, the one at the base, the other at the summit of some great elevation. We know very little of the different effect of the sun's rays at heights not usually visited, and this instrument is calculated to supply useful information on that head. So nicely will it show a decrease or increase of force in the solar heat, that the altitude of the sun may be determined by observations with it very approximately.

WIND.

76. The phenomenon of wind, which is simply air in motion, is produced by inequality of temperature in the atmosphere at different points on the surface of the earth, or in different regions of the atmospheric envelope.

If the air at a particular place is heated it will become specifically lighter; heavier air will rush towards the spot, and, occupying the lower place, will force up the heated air to the higher regions, where it will spread laterally and form an upper stratum; the horizontal currents of air are what produces the effect of wind to an observer on the surface of the earth. It is not always that wind extends to a great distance: a storm of wind and rain has been known to be raging on one side of a mountain, while the air on the other was in a state of calm.

As clouds are wafted by the wind, they will serve to show the direction of the wind in elevated portions of the air; frequently the lower clouds may be observed to be moving in one direction, and the higher in an opposite one.

77. Land and sea breezes.—The land and sea breezes of the tropics change their direction twice in the day, and the cause is obvious. When the sun shines, the surface of the land gets heated more rapidly than the sea; nor is the heat conducted far below the surface, seeing that the conducting power of the superficial layers is usually low. The air superjacent is heated from below by its contact with the surface and rises, from the cooler air over the sea flowing towards and underneath it, making the "sea breeze." After the sun has set, the surface of the earth soon radiates the heat it had received, and becomes cooler than the surface of the sea, which retains its acquired heat for a longer period. The air above the earth, being thus gradually cooled, flows outward and produces the "land breeze."

78. **Trade-winds.**—The "trade-winds" are the effect of the different degrees of heat experienced in regions within the tropics, and in those of the temperate and frigid zones. The air superincumbent over the tropical regions, which are subject to the heat of a vertical sun, will constantly receive heat from beneath by conduction, and will therefore ascend, being displaced by horizontal currents from the polar across the temperate zones. Were the earth stationary, these horizontal draughts would constitute northerly winds in the northern hemisphere near the surface, while the upper current would flow from the south, the reverse occurring south of the equator. But as the earth, accompanied by its atmosphere, revolves

from west to east in its diurnal motion, and as the rate of motion of a point on its surface increases in rapidity as the equator is neared, it is clear that air from near the poles, or from high latitudes, will bring with it less of the easterly tendency than the surface of the earth within the tropics possesses; hence it will lag, as it were, or the tropical districts will, so to speak, strike against it or leave it behind; and by composition of the two motions, the original northerly wind will appear as coming from the north-east; and wind coming from the south polar regions will be reckoned south-east: where these two currents meet, not far from the equator, is a "region of calms," across which ships can only reach by catching every fitful breeze.

The cold air from regions north and south of the tropics, by friction against the surface, partakes by degrees of the motion of the earth in its passage towards the equator; if it could be transferred suddenly from higher latitudes, its velocity would be that of the most fearful hurricane; as it is, the effect of the inrush of colder air results in the mild and constant "trade-winds," which blow constantly across those portions of the Atlantic and Pacific oceans which lie between or not far beyond the tropics.

79. **Circular storms.**—The origin of the keen

and biting east wind with us is in the region of arctic cold; the west and south-west winds, having traversed the ocean, have imbibed a portion of its warmth, hence they are usually genial. If two opposite currents meet, they frequently form a vortex and whirl over the ocean as cyclones or typhoons, travelling forward at a rate slow in comparison with the velocity of rotation. Mr. Redfield and Col. Reid have devoted much attention to the theory of circular storms, and have laid down principles to guide the navigator in escaping the hurricane by sailing away from its sphere of operation. According to those writers, hurricanes are whirlwinds on an enormous scale,—revolving storms in which the air is carried with extraordinary velocity round a calm centre or focus of variable extent, in the neighbourhood of which the force of the wind is greatest; while the hurricane is thus rushing round its focus, it advances onward, at a variable rate, from the place of its origin, along, in some districts where cyclones occur, a course of remarkable regularity. The cyclones vary in size from under fifty miles in diameter to several hundred.

The direction in which the hurricane blows is always against the course of the sun, or the hands of a watch, in the northern hemisphere, but with it in the southern; from this has been deduced a rule to tell in what direction a ship should steer to avoid the central part, near which, as has been said, the force is greatest. In the northern hemisphere, with your face to the wind, the centre will be on your right; in the southern, with your face to the wind, it will be on your left. The value of this rule has been attested by many hundred narratives of escape from the danger to which, without it, the mariner would have been exposed.

These fearful visitations are confined to the China sea, the ocean between Australia and Southern Africa, and the west of the Atlantic from the Gulf of Mexico to Newfoundland.

The revolving movement would appear to be propagated from place to place, not by the bodily transfer of the mass of air which constitutes the hurricane, but by the transmission of the rotatory motion from one portion of the air to another. An idea of the movement may be obtained by watching the water above a mill-dam when it is allowed to escape by a small hole only; a funnel-shaped eddy will be noticed, which will keep its position as long as the hole is open; the moment the hole is closed it will begin to wander, and will continue so to do for a considerable time.

It is the opinion of some eminent meteorologists, among whom is the Rev. Dr. Lloyd, that the circular movement of the air is by no means confined to violent hurricanes, but that it may be traced even in the gentlest breeze. Under his superintendence simultaneous observations have been taken for some years at the Coast Guard Stations in Ireland, and the discussion of these have led him to this conclusion, which would appear to be supported by the observations.

An arrangement was made a few years since, which lasted for some length of time, whereby the direction and force of the wind, and the state of the weather at 9 A.M. every day in the year (Sundays excepted), were transmitted by electric telegraph to Mr. Glaisher, and published, on the following morning, in the *Daily News*; the returns were sent from England, Ireland and Belgium, and the following remarks on one day's observations only, will illustrate the subject now under discussion.

November 20th, 1850, is worthy of special notice, as it is evident from the returns that the air was moving in a circle on the morning of that day; the direction of the wind in Ireland being N.W., north of France W., on the south-east coast of England and Belgium S.W., and on the south of Scotland S.E. On the outside of the circle, of which Birmingham seemed the centre, the force of the wind was very great, while in the middle of England a calm was registered by the observers; a projection of the course and force of the wind at the various points of observation, inserted in a map of England now before the author, shows clearly that the air, at this period, was describing concentric circles around a position not far removed from the centre of England. Such observations and registrations, taken extensively over the earth's surface, would be of the utmost value in making us acquainted with the phenomena of air in motion, and with the laws which regulate storms and hurricanes, which laws have only lately begun to receive the attention their importance demands.

80. Anemometers: Osler's.—Meteorological observers are expected to register at certain periods the force and direction of the wind; in a few places in England, namely Greenwich, Plymouth, the Royal Exchange, and Liverpool, this is done most accurately by means of Osler's anemometer, which will be described at length when we treat of the instruments in the Royal Observatory at Greenwich (see § 169). A board of one foot square is opposed to the wind from whatever quarter it may blow, and, by means of pencils attached in a peculiar manner, a register of the pressure per square foot, and of the direction of the wind, is traced on a sheet of paper divided into twenty-four spaces, each representing an hour, which is advanced by clock-work; so that the moment of increase, decrease, or change, can be seen at a glance.

81. **Whewell's Anemometer.**—Whewell's anemometer—described at full in § 170—registers the horizontal motion of the air during the same period; and the daily results, both of this register and the preceding, are published in the Registrar-General's weekly reports.

82. **Robinson's Anemometer.**—This anemometer, which seems likely to get into more general use than either of the others, consists, as will be seen on looking to Plate XI. fig. 1, of four hollow hemispheres or cups placed at the ends of two horizontal bars crossing at right angles and supported on a vertical axis which revolves freely. Dr. Robinson, who invented this instrument, found that the velocity of the cups when made to revolve by the wind was one-third of that of the air itself. The rate is measured either by a system of index wheels set in motion by an endless screw on the axis, as shown in the figure, or by more complicated machinery for continuous registration of the velocity, such as is in use at the Kew and Oxford Observatories. In the former case the dials are made to denote either the number of revolutions of the cups, or, better, the miles and decimals of a mile passed over by the air since the

last observation and setting of the index; the present velocity, if it is required, can be got by noting the space passed over in a certain limited time, as a minute or two. In the latter kinds of this instrument an index or a pencil is moved from one side to the other of a sheet of paper which is carried along endways by clock-work, so that the inclined line traced (by photography at Oxford, by contact of a metal pencil at Kew) shows the velocity of the wind at every moment of the day. In the very ingenious arrangement at the Kew Observatory, devised by Mr. R. Beckley, the pencil is a strip of brass placed spirally round a cylinder made to revolve by connexion with the axis of the anemometer in contact with another cylinder covered by a sheet of the registering paper and set in motion on an axis parallel to that of the last by clock-work; as the first cylinder goes round, different parts of the spiral pencil touch the paper in succession till the mark reaches the extreme end, when the pencil again comes into play at its first point. This construction is fully described, with plates, in the 'Report of the British Association for 1858,' while an account of the one at the Radcliffe Observatory, Oxford, is given with the meteorological observations taken there in 1856.

83. **Lind's Wind-gauge.**—If such complete methods were in more general use, we should no doubt

be much assisted in arriving at a knowledge of the laws regulating the movement of the air; meanwhile, although we cannot accomplish as much as we desire, it behoves us to contribute as much as lies in our power;—a useful and available wind-gauge for private observers is Lind's; and on its indications are founded the reports of the force of the horizontal movement of the air, which are recorded by those who are not provided with more elaborate apparatus. Plate II. fig. 5.

Lind's wind-gauge consists of a glass tube, *c*, about half an inch in diameter, of a siphon-like form, one end, *d*, being again bent at right angles to the general direction of the tube, so as to present a horizontal opening to the wind. The tube is half-filled with water, and the pressure of the wind on that portion directed towards it will drive the water up the other leg. A scale is attached, by which the force of the wind is ascertained; and the whole turns freely on a vertical axis, P, so that the mouth may always be towards the quarter from whence the wind blows; or a vane may be fixed to it above in the same plane as the tube, which will ensure the mouth being directing towards the wind.

A new form of this wind-gauge has been constructed by Sir W. Snow Harris, in which the after tube is made of a smaller bore than the other, and a plumb-line is made to hang within the frame of the instrument, protected by a plate of glass, to show exactly when it is in a vertical position.

The Table below shows the pressure per square foot for the indications of the scale.

Not having a convenient place to fix this instrument,—for it should be far above the interference of buildings or trees, and screwed into a firm support,—I generally estimate the force of the wind from the knowledge gained by its occasional use. Many observers do so without any reference to the wind-gauge at all; and from following the directions in the Table subjoined, they cannot be far out. A calm is universally represented by 0; a hurricane or violent gale by 6.

Table showing the force of the wind on a square foot for different heights of the column of water in "Lind's Wind-gauge."

Inches.	Force in lbs.	Common designation of such a wind.
6	31·75	A hurricane.
5	26·04	A very great storm.
4	20·83	A great storm.
3	15·62	A storm.
2	10·42	A very high wind.
1	5·21	A high wind.
·5	2·6	A brisk gale.
·1	·52	A fresh breeze.
·05	·26	A pleasant wind.
0·	0·	A calm.

84. Graphic delineation of results of wind ob-

servations.—The resultant of all the forces of the winds, combined with their directions, may be graphically shown for any definite period of time, as one month. To illustrate the method, I refer to my own observations, taken three times daily during the month of October 1848; it will be sufficient to take eight out of the thirty-two points of the compass: in the following Table are arranged the sum of the forces of the different winds registered during the month, from which figure 6, Plate II., is constructed.

	N.	N.E	E.	S.E.	S.	S.W.	W.	N.W.
Force..	2	5	0	1	10·7	29·7	3	3

Describe a circle, and divide it into eight equal portions, which mark with the names of the points of the compass used in the Table.

Towards the south, set off from a scale of equal parts, two divisions, seeing that that amount of direction southward is due to the northerly wind. On the other lines of direction set off, from the same scale, parts equal to the amounts recorded above, but in opposite directions to those of the winds; join the points thus obtained, and the general direction of the wind for the month will be seen: in this case it is evident that the whole mass of air was moved, during the month, horizontally towards the north-east. If this be done for every month of the year, the prevalent winds, both in amount and direction, can be compared with little trouble, and far more readily than by the inspection of the register; care must be taken that the same scale be used throughout, and if great nicety is required, the whole thirty-two points of the compass may be laid down. The same plan may be adopted for much shorter periods, if any remarkable changes take place which it may be important to register.

85. **Rules for observing.**—As observations on the wind may be easily made by those who are not provided with barometers or thermometers, and even more advantageously than by students, if their occupation leads them to be much in the open air, the following directions are given in order that such observations may become of scientific value. It is most important to remark, in addition to the intensity and direction of the wind at certain fixed hours,—

1. The time when it commences to blow from a calm, or subsides into one from a breeze.

2. The time of any remarkable or sudden change of direction.

3. The course it takes in varying, and the quarter in which it ultimately settles.

4. The existence of cross currents in the higher regions of the atmosphere, as indicated by the course taken by the higher clouds.

5. The times of setting-in of hot and cold winds, and the quarters from which they come.

6. The connexion of rainy, cloudy, or fair weather with the quarter from which the wind blows, or has blown for some time previously.

Such series of observations, continued through fixed periods in various parts of the country, would be of the utmost value in meteorology; and if the stations were very numerous, would throw light on the aërial movements, a branch of the science in which we are much in want of well-arranged and abundant registers.

2. *The Hygrometric Condition of the Air.*

86. General considerations.—The atmosphere is subject to agitations far more extensive than the swell of the ocean; waves as broad as the Atlantic itself pass over us from time to time. Independently of these atmospheric waves, hourly variations in the pressure of the air have been remarked which seem to follow a definite law. The cause of these fluctuations is variation of temperature depending upon the sun's hour-angle; and this also affects the amount of aqueous vapour, which combines with the air in raising or depressing the mercury in the barometer. Such is the tendency of aqueous vapour to rise in the air, that the atmosphere may be said, in no case, to be found in a state of absolute dryness; the supply is obtained from rivers, oceans, and the surface of the soil. It is possible to separate, experimentally, from a given quantity of air, the aqueous and gaseous atmosphere; but more important for the purpose of meteorological investigation are the means we possess of ascertaining the amount of vapour of water commingled with the air at any moment, so that, from the amount of pressure shown by the height of the barometric column, we can attribute to each atmosphere, the aqueous and the gaseous, exactly the proportion which is due to it. The experiments of Dalton and others have proved this remarkable fact, in a way which will be pointed out hereafter, viz. that, under similar circumstances as to temperature, the quantity of watery vapour existing in air will be exactly equal to what it would be in a vacuum of equal capacity; and that, if we have the means of computing the tension or elastic force of vapour in vacuo, we shall be able to determine with equal accuracy the actual tension of moisture in the air; in fact, that in either case the tension or pressure or elasticity of vapour will be the same.

87. Dew. Hoar-frost.—The capacity of air for moisture increases with the temperature; and when the limit, varying with the temperature, is attained, no more moisture will ascend, and the air is then in a state of complete saturation. If from any cause a volume of air in this condition should be suddenly cooled, a deposition of moisture succeeds; the air parts with aqueous vapour in minute particles, and, especially if it be free from agitation, these appear in the form of *dew,* which is witnessed in perfection after the removal of the heat of the sun on a still, autumnal night. The effect of a temperature below the freezing-point will be to convert the dew into *hoar-frost,* as is visible when winter approaches.

88. Rain. Hail.—Should the air part from its moisture at a distance from the earth's surface, the aqueous particles will descend at first but slowly, or not at all. By the law of aggregation, they will unite in globules; and, when their weight is sufficient to overcome atmospheric resistance, or, perhaps, when the electric state of the moisture or the air undergoes a change from some unknown cause, they will descend by gravitation towards the earth, and fall in the form of *drops of rain.*

That electricity is concerned in the production of rain is more than probable, as is indicated by the copious and heavy showers which fall during a thunder-storm. Its agency in the phenomenon of hail is undoubted; this is formed in the upper regions of the air, where the temperature is below the freezing-point; the height from which the hail-stones descend is indicated by the force and rapidity of their fall.

Evaporation and condensation of aqueous vapour tend, in various ways, to diffuse heat more equally throughout the globe. As the power of air to imbibe moisture increases with the temperature, evaporation goes on with most rapidity in warm climates, and from heat being absorbed in the process, has a cooling tendency; in the less heated regions the vapour is condensed, and its latent heat is given forth to mitigate the severity of the colder climate.

89. Condensation of the vapour in the air.—If two saturated volumes of air of unequal temperatures, and therefore of varying capacities for moisture, meet each other, their tendency will be to unite and equalize both the temperature and the moisture. The moisture, however, will always be in excess, for the two processes will not proceed at the same rate. Thus, suppose two volumes of air saturated with moisture, one of the temperature of 40° and the other of 60°, to unite, the mean temperature of the mass will be 50°. The elastic force of vapour at 40° of temperature (as will be explained more fully hereafter) is 0·247 measured as pressure in inches of mercury; at 60° it is 0·518; but at 50°, the mean of the two temperatures, the elastic force is 0·361, which is less than 0·382, the mean of the others, by 0·021; this represents the tension of that portion of aqueous vapour which would be set free. If this vapour, in its liberated state, meet with a stratum of air not saturated with moisture, it will be re-absorbed, either partially or entirely; if only partially, the remaining portion, consisting of aqueous vapour in an extremely minute state of subdivision, may be arrested in its descent and float in the atmosphere in the form of clouds.

From the preceding observations, it will appear that the atmospheric pressure, as shown by the barometer, is compounded of that of the air itself and the co-existing vapour of water, from which it is never entirely disunited. It will now be shown in what manner these two forces may be separated and the proper value assigned to each, according to the views of many eminent meteorologists.

90. Saussure's Hygrometer.—The plans adopted by observers in the last century to determine the hygrometric condition of the air were imperfect and unsatisfactory; the contraction and dilatation of some animal substance indicated changes in the amount of moisture in the air, and the register of these formed the hygrometers of De Luc and Saussure. At the present state of the science of meteorology such contrivances are only interesting historically, and on this ground simply we shall describe Saussure's hygrometer, which is represented in Plate VI. fig. 3.

The action of this instrument depends upon the longitudinal expansion of a human hair, which has been freed from all unctuosity by boiling in an alkaline solution. It is kept in equable distension by a small weight at one end, the other being fixed; the hair, passing round an axis which carries an index, on its contraction causes the index to revolve and point to certain divisions on a circular plate. Saussure determined the point of extreme dryness by placing the instrument under a receiver, in which was inserted some powerful desiccant; here the hair attained its minimum of length; this was the zero of his scale. He then introduced moistened pieces of linen which had been previously weighed, and on their withdrawal the diminished weight denoted the amount of vapour in the receiver; the temperature being noted at the same time, and also the portion of the arc moved over by the index, another point on the divided arc was secured, and thus data were obtained for arriving at an approximation—rough, it is true—to the amount of moisture existing in the air at any precise moment. The superiority of the methods now adopted to arrive at this conclusion, will be easily apprehended from the description of the hygrometers at present in use. We must, however, preface the description with a consideration of the laws to which the vapour of water is subject.

91. Tension of aqueous vapour.—The laws of the formation of vapours, and the relations existing between their elastic force and temperature, were ascertained by Dalton experimentally, and the results were published by him in 1802. The following is an account of the apparatus employed by him in his investigations.

In Plate IV. fig. 3, *a, b* is a barometer tube inverted in a vessel of mercury in the manner described in § 11. The graduated scale *s* measures the height of the mercurial column, its lower point being brought to touch the surface of the mercury in the vessel. Before the tube is inverted, its inside is moistened with the liquid whose vapour it is intended to experimentalize upon,—we shall only consider the vapour of water—which, being lighter than mercury, will rise and float upon the upper surface of the column, forming a thin film over it. The mercury will then be found to sink lower than it would do were the space above it, as in the barometer, a vacuum; and the amount of depression measured on the scale is the elastic force of the vapour of water due to the existing temperature.

Suppose the tube to be elevated further above the reservoir of mercury, but not lifted out of it, there will then be a larger space to be occupied by the aqueous vapour; as long as there is any liquid above the mercury, evaporation will go on to fill the increased space, and its elastic force (the temperature, pressure of the air, and consequently the density, being the same) will remain constant. The amount of descent of the mercury in the tube produced by the aqueous vapour, measures the *maximum density at the given temperature.*

When the liquid has entirely evaporated, if the space in which it exists be enlarged, its elastic force will diminish; if, again, it be contracted in volume, its density will become greater, until it attains the maximum density due to the given temperature; after this a further diminution of volume will reduce some portion of it to a liquid state.

The upper part of the tube is surrounded by a glass cylinder, *c d*; this is closed at both ends with a cork, admitting the tube through it. By filling this cylinder with water at different temperatures, we subject the vapour to the operation of any degree of heat we wish; the scale attached enables us to measure the depression of the mercury at each degree, and thus is formed a table of the elastic force of vapour for a range of many degrees. By using melting ice for the freezing-point, and freezing mixtures for temperatures still lower, this table has been carried to a great extent; above the boiling-point hot oil has been made use of; but for the purposes of meteorology, the table of the tension or elasticity of aqueous vapour may be confined to much narrower limits.

If the space above the liquid in the barometric tube is occupied by air (or indeed by any permanent gas), the vapour of water will follow the same laws as regulate it in vacuo, that is, the same relation will be preserved between the density, tension, and temperature; they will, in fact, be precisely the same as in vacuo, with this difference only, that, while in vacuo the quantity of vapour necessary to saturate the space above the column is formed *instantaneously*, *time* is requisite for the evaporation to go on in air.

The conclusions of Dalton have been confirmed more recently by Dr. Ure, Regnault, and others; they have been the foundation of tables showing the elastic force of vapour, as measured in inches of mercury, for all degrees of temperature. Dr. Ure's table was constructed from experiments performed by an apparatus of which the following is a description (Plate IV. fig. 4).

D L E is a siphon barometer, the leg E being closed, and the other open at D. On the admission of the mercury there will of course be an equilibrium when the column in the closed leg balances the atmospheric pressure on the surface of the mercury in the other, and the space above G will be a vacuum; a glass vessel is adapted to the outside of this portion of the tube, and rings of platinum wire on the exterior of the tube serve to mark the height of the mercury in each leg. A drop of water is now introduced into the vacant space above G, which is forthwith, in part, changed to vapour, and the vessel A is filled with water, the temperature of which, shown by the thermometer inserted therein, will give that to which the tension of the vapour is due. The elasticity of the vapour will cause the mercury to descend in the tube E, and rise in the tube D; a portion of the metal is now poured into the open tube, until its weight counterbalances the tension of the aqueous vapour, and brings the mercury to its original level at G. Let O L be the space occupied by the additional quantity of mercury; this space, accurately ascertained, will be the measure, in inches of mercury, of the elasticity of the aqueous vapour at the temperature shown by the immersed thermometer. By varying the heat of the water, by using freezing mixtures for the low temperatures, and boiling oil for the higher, Dr. Ure obtained the tension of aqueous vapour for degrees of heat ranging between 24° and 312°, the elastic force of the former being 0·17 inch, and of the latter 167 inches. Table II. (Appendix) is copied from the Greenwich observations; it shows the tension of aqueous vapour for every degree of temperature from 0° to 95°.

92. Important conclusion.—The point more particularly to be borne in mind from the preceding paragraphs is, that, at a certain temperature, a definite amount of vapour will arise by evaporation, and that amount will be constant, whether the space into which it rises be a perfect vacuum, or whether it be the open air, the only difference being that in the latter case the process will go on much more slowly.

93. The dew-point.—If a volume of air of the temperature of 35° be saturated with moisture, it will be seen, from inspecting Table II. (Appendix), that the tension of such moisture is equivalent to 0·222 inch of mercury, and it is concluded that if the air were suddenly to part with its moisture, the barometric column would fall to that extent.

Should the barometer reading be 29·886, the pressure of dry air vertical to the place of observation would be $29{\cdot}886 - {\cdot}222 = 29{\cdot}664$.

Complete saturation, however, is of comparatively rare occurrence; only, indeed, as we shall see, when the wet- and dry-bulb thermometer readings are the same, which will occur when the tension of aqueous vapour is at a maximum for the temperature of the air. In every other case the air is capable of holding in suspension a greater quantity of aqueous vapour than it contains at the instant of observation. The knowledge of the dew-point here comes to our assistance. The dew-point is that temperature at which the air would be saturated with the moisture it contains; and the abstraction of heat to the smallest amount below this point of saturation, would cause an immediate deposit of water.

Dalton's method of ascertaining the dew-point was a very simple one and easily practised; he poured water, cooled below the temperature of the atmosphere by a freezing mixture, into a glass, and marked the temperature by a thermometer inserted in it when the dew which had been deposited on the outside of the glass disappeared in the open air.

The tension of vapour for each degree of the thermometer differs, in a slight degree, as determined by different physicists; the table published in the Greenwich observations is derived from the experiments of Dalton and Ure (compared, however, with other tables), whose methods have on that account been explained.

Tension of vapour, according to various authorities, for 32°, 95°, and 122° of temperature.

°	Dalton.	Ure.	Roy. Soc.	Regnault.	Kaemtz.
32	0·200	0·200	0·186	0·18124	0·17999
95	1·59297	1·640	1·610	1·6438	1·57789
122	3·500	3·512	3·542	3·619	

94. Degree of humidity.—With a table of the tension of aqueous vapour and the dew-point, we are immediately provided with data for determining another important element, namely the degree of humidity of the air.

Suppose the temperature of the air to be 67°, and that of the dew-point to be 50°. By inspecting Table II. (Appendix), we find the tension of aqueous vapour at 67° to be ·659, at 50° it is ·373; if the air were saturated the tension would be ·659; but, in the instance supposed, the tension of aqueous vapour is only ·373; the proportion, therefore, between the amount of vapour in the air and that which would exist in it were complete saturation reached, is as 373 to 659; or considering the amount of vapour at the point of saturation as unity, $659 : 373 :: 1 : \frac{373}{659}$ or ·566, which will represent the degree of humidity of the air when its temperature is 67° and that of the dew-point 59°.

As this element gives immediately the preponderance or defect of aqueous vapour in the air, which the dew-point does not, it is one of considerable importance in the register of phenomena. The dew-point may be the same at extremely different degrees of air-temperature; and the state of the air, as regards dryness, will depend on the greater or less depression of the dew-point temperature below that of the air.

When the degrees of humidity are registered, we see at once the variation in the hygrometric state of the air, and comparisons are readily instituted.

If from the tension of aqueous vapour at the air-temperature, we subtract that at the dew-point temperature, we shall obtain the force of evaporation; thus, in the preceding example, ·659—·373 =·286.

95. **Illustrations.**—The phenomenon of the dew-point wil receive illustration from the affairs of common life.

A bottle of wine has been iced before its appearance in the dining-room—the bottle will be found covered with a coating of dew the moment it enters the room; the temperature of its contents being far below the point of saturation of the air, the watery vapour will be condensed on the surface; if it be removed, more will settle, till the wine has acquired a heat above the temperature of the dew-point.

On entering an observatory in the morning, the temperature of the air having risen very considerably above that of the preceding night when the instruments were in use, nothing is more usual than to find every instrument streaming with moisture. The instruments, having been cooled down to the night-temperature, and the day chancing to become suddenly warm, have not had time to get heated above the temperature of the dew-point, and an abundant deposition of moisture is the result.

On entering a detached room on a morning which may happen to be warm after a series of cold days and nights, the walls may be frequently found covered with moisture, which will be due to their not having acquired a temperature, since the sudden change, above that of the dew-point; the moisture will disappear as soon as they have attained a heat but just exceeding it.

96. Daniell's hygrometer.—Instruments have been invented for obtaining the dew-point by direct observation, three of which we shall describe. The first, that of the late Professor Daniell, known as Daniell's hygrometer, is a very elegant and portable instrument.

It consists of two glass balls, communicating

with each other through a bent glass tube. The ball *a*, Plate VI. fig. 1, is of black glass, about one and a quarter inch in diameter; the ball *d* is of the same size, but transparent. A small mercurial thermometer is fixed in the limb *a b*, with a pyriform bulb, which descends to the centre of the blackened ball. A portion of sulphuric ether, sufficient to fill three-fourths of the ball *a*, is introduced, and the atmospheric air having been expelled by boiling as completely as possible, the whole is hermetically sealed at *e*. The ball *d* is covered with muslin, and the apparatus is supported on a brass stand, *f g*, to which is fixed another delicate thermometer, whose readings show the air-temperature. To ascertain the temperature of the dew-point, the ether is all brought into the ball *a*, by inclining the tube; the temperature of the air is then registered, and ether is poured from a dropping tube, which fits the mouth of a small phial, on the muslin cover, *d*; the cold produced by the evaporation causes a condensation of the vapour of ether which fills the connecting tube and the ball, *d e*, and produces a rapid evaporation from the ether in *a*; as this is a cooling process, the temperature shown by the enclosed thermometer rapidly sinks. The instant that the ether within the black ball is cooled down to the temperature of the dew-point, a film of condensed vapour from the air surrounds the ball like a ring at the level of the surface of the ether within, *a i*; if the thermometer be read off at the instant this ring of dew is formed, we obtain nearly the true temperature of the dew-point, that of the air at the time being shown by the exterior thermometer, *f g*.

The ring of dew will gradually lessen as the ether within the ball is recovering its original temperature, and will finally disappear. At that instant the enclosed thermometer should be again read off, and this reading will give another approximation to the dew-point temperature; the mean of this and the former determination may be supposed to approach very near the truth.

The portability of this beautiful instrument is a great recommendation; the whole apparatus is packed in a box, which may be readily carried in the pocket. An important deficiency is the smallness of the enclosed thermometer, which will not admit of a reading by estimation of nearer than half a degree; other drawbacks to its use are the difficulty of obtaining pure ether for the experiment, and of catching the instant of the formation of the ring of dew, seeing that the thermometer has to be carefully watched and read off at the same moment that the dew appears or vanishes.

97. Regnault's hygrometer.—The dew-point may also be obtained by direct observation from Regnault's hygrometer, and it is presumed with a nearer approximation to the truth than Daniell's will supply.

Regnault's hygrometer (Plate VI. fig. 4) consists of a glass tube over which is slipped a thimble, *a, b, c,* made of silver, very thin and highly polished, 1·8 inch in depth and 0·8 inch in diameter; it is fitted tightly on the glass tube, *c d,* which is open at both ends.

The tube leads by a small aperture into the hollow upright support, *n m.* The upper opening of the tube is closed by a perforated cork (sometimes of india-rubber) through which passes a very sensitive thermometer, T, which, being much longer than any that can be used with Daniell's hygrometer, is more nearly and readily read off; its bulb descends nearly to the bottom of the silver thimble, and from the same depth rises a hollow thin glass tube, *g,* which also passes through the cork. Ether is poured into the tube as high as *p q*; the pipe, *d,* leads down the hollow upright, which is in communication with the flexible tube *o,* to an aspirator (not shown in the drawing), that is, a jar containing something less than a gallon of water; this jar is closed, except a small aperture, over which the flexible tube is made to fit air-tight; at the bottom of the jar is a stop-cock, on turning which the water will run out. The aspirator-jar is near the observer, but the instrument may be at any convenient distance, so as to be removed from the heat of the person.

When the dew-point is to be ascertained, water is allowed to run from the aspirator-jar; this will produce a current of air through the fine tube, whose mouth is at *g*, and the pipe of communication, *m n*; to reach the space above the ether, the air must pass, bubble by bubble, through it, which will produce a uniform temperature throughout the whole mass while it is subject to the agitation produced by the rapid passage of air, and the silver thimble will be of the same temperature as the ether within, the degree of which will be shown by the immersed thermometer, T. If very great nicety is required, the thermometer may be read off by a small telescope, but with care this is seldom necessary.

At the moment that the ether is cooled down to the dew-point temperature, the whole external surface of the silver thimble will be covered with a coating of moisture, and the degree shown by the thermometer at that instant of time must be marked. Let us suppose the first reading to be 46°, it is probable that as a fraction of a second was lost in the eye glancing from the silver to the thermometer, this reading is too low. In a few moments the dew disappears and the thermometer rises; but now its reading is probably too great, say 48°. The stop-cock of the aspirator is then gently opened, and a small stream of air-bubbles rises through the ether, and, by nice adjustment, the thermometer is observed steadily to read 47°·3 at the instant that the dew is formed; repeated trials are sometimes necessary, but five minutes will generally suffice for the whole operation. The inventor found three or four sufficient to determine the dew-point to the tenth of a degree.

The second tube, *a′ b′*, holds a second thermometer, T′, circumstanced in every way like that used in determining the dew-point, and its reading will give the temperature of the air at the time of observation. This additional thermometer need not necessarily be mounted in the same manner as the dew-point thermometer, as the temperature of the air may be learnt from one suspended in the usual way.

The direct method by which the dew-point may be ascertained by marking the clouding of the polished silver surface, is so great a recommendation, that nothing but cheapness is requisite to bring Regnault's hygrometer into general use. On the author's representation, Messrs. Negretti and Zambra have constructed the instrument in a simple form, and at a price not greater than is usually paid for Daniell's.

The aspirator need not be a vessel of water; at Kew two circular boards are united by leathern sides after the manner of a pair of bellows, to the bottom board is attached a weight and in the top board is a projection to which the india-rubber tube is attached; when an observation is to be taken, the boards are made to approximate by a pulley, and as the lower one is allowed to descend, a draught of air is produced, the force of which may be regulated at pleasure. This apparatus does away with the trouble of constantly supplying water to the aspirator-jar, as is necessary on the usual construction, though some aspirators are made double, so that they may be reversed when the water has run out of one compartment into the other, and thus the same quantity of water may be made serviceable for any length of time.

Regnault's hygrometer has not yet been used in the Greenwich observations; but, for many years, a long series of experiments and comparisons was undertaken with Daniell's hygrometer and the dry- and wet-bulb thermometers, which has resulted in the rejection, for general registration, of the use of Daniell's hygrometer, and in assuming as the dew-point, a deduction from simultaneous observations with the dry- and wet-bulb thermometers.

98. **Connell's hygrometer.**—Figure 5, Plate VI. is a representation of a dew-point instrument, by A. Connell, Esq., Professor of Chemistry in the University of St. Andrews. A is a small bottle made of highly-polished brass or silver; it is partially filled with ether, and the temperature is lowered by means of an exhausting syringe, D E, which causes the ether to evaporate rapidly; a thermometer, *t*, whose bulb dips into the ether, shows the temperature of the dew-point at the instant that the polished exterior of the bottle becomes clouded with moisture. G is a clamp to fix the instrument to a steady support.

99. **Dry- and wet-bulb thermometers, or Mason's hygrometer.**—The dry- and wet-bulb thermometers, known also as Mason's hygrometer in England, and as August's psychrometer on the continent, is the form of hygrometer in general use; it consists simply of two thermometers exactly alike, stationed side by side, which are presumed, under the same circumstances, to give similar indications. The dry-bulb thermometer, of course, shows the temperature of the air; the wet-bulb thermometer has its bulb sur-

rounded with muslin, and from it lead a few inches of lamp-wick, or floss-silk, into a small vessel filled with rain or distilled water.

Under general circumstances, or rather, whenever it is not saturated, the atmosphere will take up the vapour of water; the drier it is the more rapidly will evaporation proceed, and the more slowly as its condition approaches that of complete saturation. When in that state, no more moisture will rise in the air. Now, as evaporation proceeds, heat is absorbed by the conversion of the water around the wet-bulb into vapour, and the mercury in the wet-bulb thermometer will fall a greater or less number of degrees below the air-temperature, according to the dryness of the atmosphere. When the air is saturated, the readings will be the same*.

Evaporation from the wet-bulb will proceed even when the temperature is below freezing; but in this case the readings must be taken with great care, and the differences will always be small.

Plate VI. fig. 2, is a drawing of the arrangement adopted by Negretti and Zambra. The

* The greatest difference in the readings of the dry- and wet-bulb thermometers which the author registered during seven years of observation at Southampton, occurred April 19, 1854, at 3 P.M., when the dry-bulb thermometer reading was 69° and the wet-bulb 53°: diff. 16°.

thermometers are very superior instruments, with enamelled divisions on the stem, and they are fixed to a frame of plate-glass. It will be remarked that they are not very near each other; and this is important, for it may be that the air surrounding the dry-bulb would be, to a certain extent, moistened by evaporation from the water supplied to the wet-bulb. It is advised that the cup of water should even be removed several inches from the wet-bulb, so that there may be no chance of incorrectness from this source. The thermometers are kept in place by metal clips, *e*, *f*, and may be used for other purposes, if required, being removeable at will.

100. Dr. Babington's Evaporation-gauge.—The evaporation-gauge invented by Dr. Babington will be found convenient for estimating the absolute amount of evaporation taking place over a given area. It consists of a vessel containing water in which floats a glass tube poised like a hydrometer, and graduated to grains and fractions of a grain; on the top of this tube is fixed a shallow pan whose area is known; water is poured into this pan until the tube sinks to the point marked zero on the scale; as evaporation goes on, the tube rises, and the loss of weight is indicated on the scale in grains and fractions of a grain.

101. Apjohn's formula.—The reading of the

wet-bulb thermometer gives the temperature of evaporation; and an important problem to be solved is to deduce the dew-point from this record combined with the temperature of the air as indicated by the dry-bulb. Dr. Apjohn of Dublin, in 1834 and 1835, read before the Royal Irish Academy a very elaborate paper on the theory of the moist-bulb hygrometer; and from observation, experiment, and theoretical considerations, he was induced to adopt the formula which has since gone by his name.

Let f = tension of aqueous vapour at the dew-point temperature which we desire to know.

f' = the tension of vapour at the temperature of evaporation, as shown by the wet-bulb thermometer.

a = the specific heat of air.

e = the latent heat of aqueous vapour.

$(t-t')$ or d = the difference between the reading of the dry-bulb thermometer and that of the wet.

p = the pressure of the air in inches: then Apjohn's formula is

$$f=f'-\frac{48a(t-t')}{e}\times\frac{p-f'}{30};$$

or with the coefficient,

$$f=f'-\cdot01147\,(t-t')\times\frac{p-f'}{30}.$$

The following is the formula (derived from Apjohn's), as given in the Greenwich observations, which will be made use of in this work; *h* being the height of the barometer.

$$f=f'\frac{d}{88}\times\frac{h}{30}.$$

When the reading of the wet-bulb thermometer is below 32°, the formula becomes

$$f=f'\frac{d}{96}\times\frac{h}{30}.$$

102. Example of its use.—Suppose the reading of the dry-bulb thermometer to be 67° and of the wet 59°, it is required to determine the dew-point, the barometer standing at 29·000 inches. By Table II. (Appendix), we find f' the tension of aqueous vapour due to 59°=·5; then, by substitution,

$$\cdot 5-\frac{8}{88}\times\frac{29}{30}=\cdot 412 \text{ or } f.$$

The tension corresponding to ·412 in the Table is due to 53°·6, which is therefore the temperature of the dew-point required.

103. Other formulæ.—Other physicists have given separate formulæ. One, practically equivalent to Apjohn's and originally put forward by August, has been modified by Regnault, thus:—

$$n=f'-\frac{0\cdot 429(t-t')}{610-t'}h,$$

where n is the tension due to the dew-point, t the temperature of the air, and t' that of evaporation in degrees Centigrade, f' and h as before.

Take the same example, and changing 67° Fahrenheit into 19°·44 Centigrade, and 59° F. into 15° C., we have, by substitution,

$$\cdot5 - \frac{\cdot429 \times 4\cdot44}{610 - 15} \times 29 = \cdot407$$

tension due to 53°·3, the dew-point.

Burg, from observation, has given (p being the height of the barometer)

$$f = f' - \cdot0004528(t - t')p,$$

which will give by substitution,

$$\cdot5 - \cdot058 = \cdot442$$

tension due to 55°·5, the dew-point; and Bohnenberger, also from observation,—

$$f = f' - \cdot0003962(t - t')p,$$

which, by substitution, gives

$$\cdot5 - \cdot051 = \cdot449$$

tension due to 56°, the dew-point.

By Glaisher's factors, hereafter to be explained, 67 − (67 − 59)1·8 = 52·6, the dew-point; while the mean of the other four results is 54·6.

By comparing these results, it will be seen that there is a certain amount of discrepancy between them; nor is this surprising, when we consider that the fundamental quantities which enter into the calculation, viz. the specific heat of air and the latent heat of aqueous vapour, are not determined with absolute certainty.

104. Glaisher's factors.—This circumstance, and other considerations, led Mr. Glaisher to undertake his elaborate series of comparisons with Daniell's hygrometer, and the dry- and wet-bulb thermometers. In practice, he at times experienced difficulties in the use of the former, and not unfrequently found that the simultaneous results of the dew-point, as found from Daniell's hygrometer and the dry- and wet-bulb thermometers, were discordant, and on investigating the causes he considers that the error rested solely with Daniell's hygrometer. The times at which these discordances existed were in those particular states of the air when great dryness was prevalent, and the depression of the temperature of the dew-point below that of the air was great, and a long time elapsed after the dropping of ether on the white ball, before dew was deposited on the black ball. Such would require the long continuance of the observer near the instrument, which would necessarily influence both the hygrometrical state and the temperature of the air around the instrument; and this would be especially the case if the observer were short-sighted, and obliged to approach the instrument very nearly. He makes the following objections to the use of this hygrometer:—

"Supposing the inclosed thermometer to be one of extreme delicacy, which it is not, it would then indicate the temperature of the portion of ether only with which its bulb was in contact, and which portion may be very different indeed from that part of the outside of the glass upon which the dew is deposited. And if the ether be dropped very slowly upon the white bulb, so that evaporation should proceed very slowly, the evil of long-continued watching is required; and if more quickly, then the different layers of the enclosed ether will be of different temperatures. It must also be recollected that the situation of the black ball upon which the deposit of dew takes place, is not very far from the white ball, and in cases where large quantities of ether are necessary, such must influence materially the hygrometric state of the air in the space included betwee both bulbs."

In consequence of these sources of error in the use of Daniell's hygrometer, together with its expense in use and trouble of using, Mr. Glaisher made many attempts, by different combinations of the results derived from the observations of the dry- and wet-bulb thermometers, to deduce the temperature of the dew-point from them; and at last he found that the difference between the temperatures of the air and evaporation was constant at the same temperature; but that this value was different with every different temperature.

He then collected all the simultaneous observations, amounting to many thousands, which had been made at Greenwich from the year 1841 to 1854, and combined with them some observations taken in India for the higher temperatures, and some at low temperatures that were taken at Toronto. From these he deduced the following Table, the use of which is very simple. From the reading of the dry-bulb subtract the reading of the wet, multiply the remainder by the factor standing in the Table opposite the reading of the dry-bulb thermometer, subtract the product from the dry-bulb reading, and the result will be the temperature of the dew-point.

Thus, if the dry-bulb thermometer reads 68° and the wet-bulb 63°, we take the factor 1·7 from the Table (opposite to 68°), and multiply it into the difference of the two readings, $(68-63)\times 1{\cdot}7=8{\cdot}5$, and subtract the quantity from the reading of the dry-bulb; so that we get 59·5 for the temperature of the dew-point when the readings of the two thermometers are as above.

Table of Factors.

Reading of the Dry-bulb Thermometer.	Factor.	Reading of the Dry-bulb Thermometer.	Factor.	Reading of the Dry-bulb Thermometer.	Factor.
20°	8·1	44°	2·1	68°	1·7
21	7·8	45	2·1	69	1·7
22	7·6	46	2·1	70	1·7
23	7·2	47	2·1	71	1·7
24	6·9	48	2·1	72	1·7
25	6·5	49	2·0	73	1·7
26	6·0	50	2·0	74	1·7
27	5·6	51	2·0	75	1·7
28	5·1	52	2·0	76	1·7
29	4·6	53	2·0	77	1·7
30	4·1	54	1·9	78	1·6
31	3·7	55	1·9	79	1·6
32	3·3	56	1·9	80	1·6
33	3·0	57	1·9	81	1·6
34	2·7	58	1·9	82	1·6
35	2·6	59	1·8	83	1·6
36	2·5	60	1·8	84	1·6
37	2·4	61	1·8	85	1·6
38	2·3	62	1·8	86	1·6
39	2·3	63	1·8	87	1·6
40	2·2	64	1·8	88	1·6
41	2·2	65	1·8	89	1·6
42	2·2	66	1·8	90	1·6
43	2·2	67	1·8	91	1·6

Many meteorologists, however, prefer to calculate the dew-point by Apjohn's formula, and with a little practice, aided by the table of the elasticity of aqueous vapour, it may be done almost as readily as by the factors given above. Had the series of comparisons been instituted between the dry- and wet-bulb thermometers and Regnault's dew-point instrument, the result would have been far more satisfactory. Such a series has yet to be undertaken for the advantage of observers in general, through very considerable ranges of temperature, humidity, and pressure, and it would either correct or confirm the truth of Mr. Glaisher's determinations.

105. Other deductions.—Having ascertained the dew-point, upon which every other deduction depends, we are prepared to determine other very important particulars respecting the condition of the air; we shall first explain the method of ascertaining the weight of a cubic foot of air of any density and temperature.

106. Weight of a cubic foot of dry air.—It was experimentally determined by M. Gay-Lussac that air expands $\frac{1}{480}$th of its bulk for every addition of 1° of heat; inasmuch as it was found to expand equally, with equal increments of heat, from the freezing- to the boiling-point to the amount of $\frac{3}{8}$ths of its bulk.

The later investigations of M. Regnault have shown that this expansion is somewhat in excess; his researches have led physicists to adopt $\frac{1}{491}$, or more nearly still $\frac{1}{491 \cdot 2}$, as the rate of expansion for each degree of Fahrenheit. (See § 21.)

The whole of the calculations in the Greenwich Tables are, however, founded on the earlier determination; while the determination of the weight of a mass of dry air, which shall be one cubic foot in volume under a pressure of 30 inches, and at a temperature of 32°, is derived from the experiments of Sir George Shuckburgh, Biot and Thénard, which are not in accordance with those of Regnault.

Regnault gives the weight of 100 cubic inches (the French measures being reduced to English) of dry air deprived of carbonic acid gas = 32·58684 grains; the barometric pressure being 29·92 inches, and the temperature the zero of the Centigrade scale, or 32° Fahrenheit; hence the weight of a cubic foot of such air is 563·1 grains. The weight therefore due to 30 inches of barometric pressure will be

$$563{\cdot}1 \times \frac{30}{29{\cdot}92} = 564{\cdot}6 \text{ grains.}$$

The weight of a cubic foot of dry air, at the same pressure and temperature, adopted in the Greenwich observations, is 563 grains.

Taking a cubic foot of dry air at a pressure of 30 inches and a temperature of 32° as unity, a simple proportion will give the space it will occupy at any given temperature—say 44°.

We have seen that the expansion of volume is $\frac{1}{491}$ for every degree of heat; required the expansion for 44—32 or 12°; $1° : 12° :: \frac{1}{491} : \frac{12}{491}$ or 0·0244, so that a cubic foot of air increases to

1·0244 feet between 32° and 44°; but the weight of the cubic foot of air originally was 564·6 grains; hence, the weight varying inversely as the volume, 1·0244 ft. : 1 ft. : : 564·6 grains : 551·1 grains, the weight of the same volume after expansion.

In the "Table showing the weight in grains of a cubic foot of dry air under the pressure of 30 inches of mercury for every degree of temperature from 0° to 90°," published in the Greenwich Meteorological Observations, the weight of a cubic foot of air at 44° is given as 549·27.

Two other tables given in the Greenwich Meteorological Observations, as being the foundation, in union with the former just explained, of valuable calculations, may be alluded to. The first is entitled "A table showing the enlargement which a volume of dry air receives, when saturated with vapour, under the pressure of 30 inches of mercury," from 0° to 90° of temperature.

If a cubic foot of dry air of known elasticity be mixed with a cubic foot of vapour, also of known elasticity, and if the mixture be compressed into the space of one cubic foot, the elasticity of the mixture will be the sum of the two elasticities of the air and vapour; or, if it be allowed to expand till its elasticity is equal to that of the unmixed air, it will occupy a larger volume in the proportion of the sum of the two elasticities to the elasticity of the air alone. Now, from Table II. (Appendix), we know the elastic force of vapour for every degree of temperature; let it be required to find the space occupied by a cubic foot of dry air, and such an amount of aqueous vapour as will ascend into it at the temperature of 44°.

Tension of aqueous vapour at 44°=	0·288	inches.
Tension of dry air ,,	30·000	,,
Tension of both combined	30·288	,,

Then as the spaces occupied are inversely as the tension or elasticity,

30 : 30·288 : : 1 cubic foot : 1·096 cubic feet,

the space occupied by the mixture of the aqueous and gaseous fluids.

The Greenwich formula will lead to precisely the same result.

Let p=the atmospheric pressure, as measured by the inches of mercury in the barometer.

E_t=the elasticity of vapour at temperature t.

n=the bulk of a certain quantity of air when dry, at the given temperature t, and under the pressure p.

n'=the bulk of the same quantity of air when saturated with vapour at the same temperature t, and under the same pressure p.

Then since the elasticity varies inversely as the volume, the temperature remaining the same, that portion of the elastic force p which depends on the air only which occupies the space n' is

$$p\times\frac{n'}{n},$$

and this, together with E_t, must make up the atmospheric pressure,

or $$p=p\frac{n}{n'}+E_t,$$

or $$\frac{n}{n'}=\frac{p-E}{p}=\left(1-\frac{E_t}{p}\right),$$

or $$n'=\frac{n}{1-\frac{E_t}{p}}.$$

By substituting the values from the above example in this expression, we obtain

$$n'=\frac{1}{1-\frac{\cdot 288}{30}}=1\cdot 096 \text{ as before.}$$

From calculations of this kind the quantities tabulated have been obtained for each degree, but as the Table is not absolutely necessary for hygrometrical deductions, it is not inserted in this work.

107. Weight of a cubic foot of vapour.—The next Table demanding explanation gives "The weight in grains of a cubic foot of vapour under the pressure of 30 inches of mercury for every degree of temperature from 0° to 90°."

As vapours expand by the increase of temperature by the same law as permanently elastic fluids, and undergo a change of volume proportional to the change of pressure, and as air expands $\frac{3}{8}$ths of its bulk* from 32° to 212°, its expansion being uniform between these points;—

Therefore if the weight of a cubic foot of vapour under the pressure equivalent to 30 inches of mercury, and at a temperature of 212°, be called W, and the weight, expressed in the same denomination of an equal volume of vapour at the temperature t, be called W′, and if E_t be the elasticity of vapour at the temperature t, then (the expansion of dry air from 32° to 212° being 0·375 or 0·002083 for each degree of temperature)

$$W' = \frac{1{\cdot}375 \times W \times E_t}{30(1 + {\cdot}002083\overline{t^\circ - 32^\circ})}.$$

Now Gay-Lussac has also determined that a cubic inch of vapour at 212° weighs 0·149176 grains under the pressure of 29·922 inches of mercury; and consequently a cubic foot of vapour, under the same circumstances, weighs 0·149176 ×1728=257·776 grains; and under a pressure of 30 inches it weighs

$$\frac{30}{29{\cdot}922} \times 257{\cdot}776 = 258{\cdot}448 \text{ grains.}$$

* This is the ratio of expansion applied in the Greenwich Tables; see, however, § 22.

Therefore, substituting this weight for W, the formula becomes

$$W' = \frac{1{\cdot}375 \times 258{\cdot}448 \times E_t}{30(1 + {\cdot}002083 \times \overline{t^\circ - 32^\circ})};$$

from a like formula but with Regnault's elements is calculated Table III. (Appendix), taken from Glaisher's Hygrometrical Tables, from which may be obtained the weight of vapour in a cubic foot of air saturated with moisture, at any temperature.

If the temperature t, denoting that of the dew-point, be also that of the air, which will only be the case when the dry-bulb and wet-bulb thermometers read alike, this Table would give by inspection the weight of vapour in a cubic foot of air, the argument being the temperature; but when, as is usually the case, the dew-point temperature is below that of the air, the vapour in the air will have expanded, and its density will have diminished in the same ratio as air. On this fact is founded Table IV. (Appendix), which is thus used.

The dew-point temperature being known, enter Table III. with it, and take out the corresponding weight (in grains) of a cubic foot of air. Entering Table IV. with the difference between the air-temperature and that of the dew-point, take out the factor opposite the number of degrees; and this, multiplied into the quantity extracted from Table III., will give the weight of vapour in a cubic foot of air at the temperature shown by the dry-bulb thermometer.

Example: let the dry-bulb read 50° and the wet-bulb 47°; the barometer standing at 29 inches.

The tension of vapour at the dew-point, from Apjohn's formula, will be

$$\cdot 323 - \frac{3}{88} \times \frac{29}{30} = \cdot 290$$

corresponding to dew-point 44°. Enter Table III. with 44°; and opposite will be found 3·3 grains; with 6°, the difference between the air-temperature and the dew-point, enter Table IV. and take out the factor ·988; then 3·3 × ·988 = 3·26, which will be the weight (in grains) of vapour in a cubic foot of air at a temperature of 50° when the dew-point temperature is 44°.

Another formula of great simplicity will give a near approximation:—let f = elastic force of vapour at the dew-point; t the air-temperature; the weight (in grains) of moisture in a cubic foot of air will be $\frac{5656\cdot 2}{448+t}f$ nearly;—by substitution, from the above example $\frac{5656\cdot 2}{448+50} \times \cdot 288 = 3\cdot 27$.

108. Amount required for saturation.—Table III. will enable us to ascertain how much more moisture would be requisite completely to saturate the air. Entering it with the air-temperature 50°, we find opposite to that number 4·1 grains, which would be the weight of vapour in the air if it were saturated with moisture; but we have found that, with a dew-point at 44°, it holds in suspension only 3·26 grains; therefore, in this case, 4·1—3·26, or 0·84 grain additional, would be absorbed before complete saturation would be attained.

109. Weight of a cubic foot of moist air.—This quantity we may obtain readily from Table V.; if the temperature of the air and the dew-point be alike, the quantity ranging with the temperature will be the weight of the air, seeing that in this case it will be saturated with moisture. But if, as is usually the case, the air-temperature be above that of the dew-point, enter the Table with the air-temperature and take out the quantity under the heading "Excess:" this, added to the weight of a cubic foot of saturated air, will give the weight of a cubic foot of dry air; the same quantity, multiplied by the degree of humidity supposed to have been determined previously, must be subtracted from the weight of a cubic foot of dry air, and the result will be the weight of a cubic foot of air of the given temperature and humidity; this last result, multiplied by $\frac{\textit{height of barometer}}{30}$, will be the true weight of the air under the existing pressure.

Taking the last example; it is required to find

the weight of a cubic foot of air under the circumstances there recorded.

The degree of humidity will be $\frac{\cdot288}{\cdot361}=\cdot8$. Entering Table V. with 50°, under "Excess" we find 2·22, which added to the number ranging with it, 544·4, will give 546·82, the weight of a cubic foot of dry air; but $2\cdot22\times\cdot8=1\cdot77$, the excess of such weight above the weight of a cubic foot of air of the existing temperature and humidity, which will therefore be $546\cdot82-1\cdot77=545\cdot05$. Finally, $545\cdot05\times\frac{29}{30}=526\cdot66$ will be the weight of a cubic foot of air at a temperature of 50°, of the degree of humidity ·8, and under a barometric pressure of 29 inches.

These are all the deductions which are practised at Greenwich; the formulæ and tables are for the most part drawn from Glaisher's Hygrometrical Tables and the Greenwich "Magnetical and Meteorological Observations," which are published yearly. The author, through the Council of the Royal Astronomical Society, has obtained for many years past a grant of these portly quarto volumes, which have been most valuable as books of reference, and he gratefully acknowledges his obligation both to the Society and the Government —to the one for their recommendation, and to the other for their liberality; he trusts that his endeavour to explain and simplify the processes employed will be found useful to a large class of persons who may never have an opportunity of consulting the records of the Greenwich meteorological observations.

It may not at first be supposed that it is possible for the air, when saturated with moisture, to hold in solution a less amount of aqueous vapour than when, the temperature being higher, it is considerably removed from the point of saturation; this will, however, be the fact whenever the temperature of the dew-point in the latter case is higher than in the former. Let us investigate the amount of moisture in a cubic foot of air when the dry-bulb thermometer reads 63° and the wet 58°; and also when the dry-bulb reading is 40° and the wet-bulb 39°.

	Dry-bulb 63°. Wet-bulb 58°.	Dry-bulb 40°. Wet-bulb 39°.
Degree of humidity	·72	·92
Dew-point	53·8	37·7
Weight of vapour in a cubic foot ...	4·6	2·6

The fact above stated will be realized when it is considered that when the dew-point is 53°·8, the air will hold in solution a greater amount of aqueous vapour than when its temperature is 37°·7. It would seem undisputed, however, that with regard to the effect of the air viewed in relation to health, the degree of humidity, rather than the absolute amount of moisture in a given quantity of air, will be the matter most worthy of the attention of the medical man. To assist in arriving at a conclusion on this point without the labour of reduction, Table VI. (Appendix) has been inserted, which will be useful in a sanitary point of view, and also for agricultural operations. It gives by inspection the degree of humidity of the air for the general range of temperature in this country, the argument being the difference between the readings of the dry- and wet-bulb thermometers for every degree of temperature from 32° to 79°, the determinations corresponding with those resulting from the dew-point as found by Apjohn's formula.

110. Welsh's Sliding-rule.—Mr. Welsh, late of the Kew Observatory, invented a sliding-rule* for facilitating hygrometrical calculations, which has been described at full in the Report of the British Association for 1851, p. 42. By means of this instrument, the use of which may be learnt in half an hour, we may obtain, with little trouble and with sufficient accuracy, the dew-point, the tension of aqueous vapour, the degree of humidity, and the weight of vapour in a cubic foot of air, the readings of the dry-bulb and wet-bulb thermometers being known.

The author has tested this useful instrument at

* Sold by Adie, Fleet Street.

various ranges of difference between the dry- and wet-bulb, and has found it beautifully accurate, the quantities agreeing with those deduced from the Greenwich formulæ to the third decimal. By those who are not ready at calculation, it will be found most useful, and those who are will obtain a result in five minutes which in any of the usual methods would occupy a much longer time.

111. Glaisher's Tables.—Mr. Glaisher's "Hygrometrical Tables*" give all the results explained in this work by inspection, when the difference between the dry- and wet-bulb thermometers is void of fractions of degrees; additional labour is, however, requisite when decimals intervene. The observations taken at upwards of fifty stations in Great Britain, and forwarded monthly to Mr. Glaisher as Secretary of the British Meteorological Society, are all reduced by the aid of these Tables, to bring them up to the same standard, and to afford the means of direct comparison among themselves, and with the atmospheric conditions registered at Greenwich, the head-quarters of meteorological science.

112. Deductions worked out in full.—The formulæ and tables explained in this part of the work will be found sufficient to perform all the

* Published by Taylor and Francis, Red Lion Court; price 2s. 6d.

deductions which are ever required relative to the hygrometric state of the air; they amount to six, the first four being the most important. An example will, at one point of view, illustrate the manner by which all these important particulars may be derived from the simple record of the temperature of the air and that of evaporation.

Suppose the reading of the dry-bulb to be 56°, of the wet-bulb 50°, the barometer reading 29·900 inches; required,—

α. The tension of aqueous vapour.—Table II. (Appendix), and Apjohn's formula, page 146,—

$$f' - \frac{d}{88} \times \frac{h}{30} = f \text{ or } \cdot 361 - \frac{6}{88} \times \frac{29\cdot 9}{30} = \cdot 293.$$

β. The dew-point corresponding to this tension of aqueous vapour in Table II. (Appendix) is 44·5.

By Glaisher's factors, page 152, these two deductions result thus:

$$56 - (56 - 50)1\cdot 9 = 56 - 11\cdot 4 = 44\cdot 6,$$

the dew-point; whence by Table II. the tension of aqueous vapour = ·294.

γ. The degree of humidity,

$$= \frac{\text{tension at dew-point}}{\text{tension at } 56^\circ} = \frac{\cdot 294}{\cdot 449} = \cdot 654.$$

δ. The weight of vapour in a cubic foot of air;

Table III. gives what it would be if the air were saturated, viz. 3·3 grs., which, multiplied by the factor opposite 11°·5 (the difference between the temperature of the air and the dew-point) in Table IV,. will be $3·3 \times ·977 = 3·22$ grains, or by the second formula, page 160,

$$\frac{5656·2}{448+56} \times ·294 = 3·29.$$

ε. The amount of vapour required to saturate the air:—the quantity of vapour held in solution by air at 56°, when saturated, = 5·; ∴ $5· - 3·22 = 1·78$, the amount required.

ζ. Table V. will give the weight of a cubic foot of air in grains: thus (see page 161),

$$\{537·5 + 2·95 - (2·95 \times ·654)\} \frac{29·9}{30} = 536·7.$$

113. Diurnal range of temperature of evaporation and the dew-point.—In the Philosophical Transactions, Part I. 1848, Mr. Glaisher has given Tables of the diurnal range of the temperature of evaporation as shown by the wet-bulb thermometer, and also of that of the dew-point; by which, from one daily observation of either, the monthly mean may be deduced, as in the case of the monthly mean temperature explained in § 48.

Corrections to be applied to the monthly mean readings of the wet-bulb thermometer placed at the height of 4 feet above the soil, at any hour, to deduce the true mean temperature of evaporation for the month from the observations taken at that hour.

Local mean time.	Jan.	Feb.	March.	April.	May.	June.
h	°	°	°	°	°	°
12 Midn.	+0·7	+1·2	+2·1	+2·9	+3·8	+3·5
1 A.M.	+0·7	+1·4	+2·0	+3·1	+3·1	+4·2
2	+1·0	+1·6	+2·2	+3·4	+4·2	+4·8
3	+1·1	+1·6	+2·4	+3·8	+4·4	+5·3
4	+1·4	+1·8	+2·7	+4·1	+4·4	+5·8
5	+1·6	+1·8	+2·8	+4·3	+4·2	+5·3
6	+1·7	+1·9	+2·7	+3·9	+2·9	+3·5
7	+1·7	+1·7	+2·6	+2·7	+1·2	+1·2
8	+1·3	+1·3	+2·0	+1·0	+0·1	−0·7
9	+0·8	+0·7	+0·3	−1·0	−1·7	−2·1
10	0·0	−0·3	−1·3	−2·3	−3·0	−3·1
11 A.M.	−1·4	−1·6	−2·5	−3·6	−3·8	−3·4
12 Noon.	−2·1	−2·5	−3·6	−4·3	−4·2	−4·0
1 P.M.	−2·5	−3·0	−4·0	−4·8	−4·4	−4·2
2	−2·4	−2·9	−3·9	−4·8	−4·3	−4·4
3	−1·8	−2·4	−3·6	−4·5	−4·0	−4·6
4	−1·2	−2·1	−2·8	−3·9	−4·4	−4·7
5	−0·6	−1·4	−1·9	−3·0	−2·6	−3·6
6	−0·2	−0·7	−1·0	−1·8	−1·4	−2·7
7	−0·2	0·0	−0·3	−0·4	−0·4	−1·5
8	−0·1	+0·2	+0·6	+0·5	+0·8	0·0
9	+0·2	+0·5	+1·2	+1·2	+1·4	+0·7
10	+0·3	+0·8	+1·6	+1·8	+2·3	+1·7
11 P.M.	+0·4	+1·0	+1·8	+2·4	+3·0	+2·6

Table (*continued*).

Local mean time.	July.	Aug.	Sept.	Oct.	Nov.	Dec.
h	°	°	°	°	°	°
12 Midn.	+3·1	+2·6	+2·2	+1·9	+1·3	+0·5
1 A.M.	+3·2	+2·9	+2·8	+1·8	+1·4	+0·7
2	+3·5	+3·3	+3·7	+2·1	+1·6	+0·8
3	+3·6	+3·4	+4·1	+2·3	+1·6	+0·9
4	+3·6	+3·5	+4·5	+2·4	+1·6	+1·0
5	+3·4	+3·7	+4·1	+2·3	+1·5	+0·9
6	+2·3	+3·0	+3·2	+2·0	+1·4	+0·8
7	+1·0	+1·2	+2·1	+1·5	+1·1	+1·0
8	−0·6	−0·3	+0·9	+0·4	+0·5	+0·9
9	−1·6	−1·7	−0·8	−0·7	+0·2	+0·6
10	−2·6	−2·6	−2·5	−2·0	−0·4	0·0
11 A.M.	−3·2	−3·6	−3·6	−3·0	−1·5	−0·8
12 Noon.	−3·5	−3·9	−4·3	−3·7	−2·3	−1·4
1 P.M.	−3·6	−4·3	−4·4	−3·7	−2·5	−1·6
2	−3·6	−4·5	−3·9	−3·0	−2·6	−1·5
3	−3·4	−3·9	−3·1	−1·9	−2·3	−1·3
4	−3·3	−3·0	−2·6	−1·2	−1·6	−1·0
5	−2·6	−1·3	−2·4	−0·9	−0·9	−0·6
6	−1·9	−0·5	−1·9	−0·7	−0·3	−0·6
7	−0·8	+0·3	−0·9	−0·1	−0·1	−0·4
8	+0·2	+1·0	+0·1	+1·2	+0·3	−0·2
9	+1·1	+1·4	+0·4	+0·6	+0·7	+0·1
10	+2·0	+1·5	+1·0	+1·0	+0·9	+0·2
11 P.M.	+2·5	+2·1	+1·5	+1·3	+1·1	+0·5

The next Table gives the corrections for the dew-point for a few hours only out of the twenty-four, those which are most generally chosen for hours of observation.

Corrections to be applied to the monthly mean reading of the temperature of the dew-point, 4 feet above the soil, to deduce the true mean temperature of the dew-point for the month from observations taken at the hours 3 A.M., 9 A.M., 3 P.M., and 9 P.M.

	Jan.	Feb.	March.	April.	May.	June.
3 A.M.	+1·1	+2·2	+1·2	+1·7	+2·9	+3·6
9 A.M.	+0·3	+0·2	+0·1	−1·0	−1·6	−1·5
3 P.M.	−0·8	−1·7	−1·8	−1·9	−2·4	−2·5
9 P.M.	−0·5	−0·2	+0·2	+0·8	+0·9	+0·7

	July.	August.	Sept.	Oct.	Nov.	Dec.
3 A.M.	+1·9	+1·8	+3·1	+1·0	+1·4	+0·8
9 A.M.	−0·6	−1·4	−0·8	−0·3	+0·1	+0·2
3 P.M.	−1·6	−2·2	−2·0	0·	−1·2	−0·6
9 P.M.	+0·6	+0·4	+0·1	+0·1	+0·5	+0·1

114. Observations in the higher regions of the atmosphere.—Observations taken at a much greater height than 4 feet above the surface of the earth would be most valuable for the promotion of meteorological investigations, for we are very little acquainted with the degree of humidity of the higher strata of the air. From observation some few facts are rendered apparent; it is found, for instance, that in the hottest hours of the day, when evaporation goes on most rapidly, the air at a short distance from the surface of the earth is relatively drier than during the cooler hours; it follows that much of the moisture must ascend into the air and be united with the upper strata, seeing that it does not increase the degree of humidity of the lower.

With the object of learning somewhat of the conditions of the upper regions of the air, the Kew Committee of the British Association for the Advancement of Science projected four balloon ascents in the year 1852, the results of which were published in the 'Philosophical Transactions' for 1853. The observations, thermometric, hygrometric, and barometric, were taken by John Welsh, Esq., a gentleman of great experience, accustomed to the practice of meteorology at the Kew Observatory, and they form a valuable scientific record.

Simultaneous observations during the time of the ascent, as well as some hours before and after, were undertaken in various parts of England, Ireland, and France, and every care was taken to determine the state of the air at the level of the sea, by combining and reducing those near the point of ascent, the Vauxhall Gardens, and below the course of the balloon. Regnault's hygrometer was occasionally used in the ascent, but the degree of humidity and the dew-point were in general deduced from the readings of the dry- and wet-bulb thermometers, by Apjohn's formula, and Dalton's Table of the tension of aqueous vapour.

The following register of meteorological observations is selected from the account of the last ascent, Nov. 10, 1852, when the highest point reached, 22,930 feet, far exceeded that of any of the three previous ascents. The mean height of the barometer at 120 feet above the level of the sea during the time of the ascent was 29·978; the temperature of the air was concluded to be, at the same time and level, 49·2.

Height above the sea in feet.	Barometer.	Dry-bulb.	Wet-bulb.	Tension of aqueous vapour.	Dew-point.	Relative humidity.
5,880	24·17	34·7	33·4	·197	31·7	·90
7,070	23·12	34·7	31·2	·167	27·1	·76
8,420	21·97	32·5	31·4	·187	30·2	·92
10,630	20·17	26·9	24·2	·132	20·6	·79
13,860	17·75	15·9	12·4	·076	5·8	·68
15,820	16·44	12·8	8·1	·057	− 1·8	·58
18,700	14·60	2·2	−2·1	·035	−14·0	·53
22,640	12·4	−8·9				

The following are Mr. Welsh's remarks on the ascent of November 10th:—

"The humidity, as in all the previous series, increased from the earth to the first cloud, which was at a low elevation and of but little density; upon leaving it, at about 1900 feet, a slight depression took place. Immediately above this low cloud a different current of air existed, shortly after entering which the humidity again increased, until, in the second cloud, it became nearly complete; the decrease, after leaving the cloud at 5000 feet, becoming rapid, and attaining a minimum at 6500 feet. A second well-defined maximum was reached at 8300 feet, followed at 10,000 feet by a secondary minimum. The humidity diminished on the whole till about 15,800 feet, when a sudden increase commenced, which continued from 16,500 to 17,600 feet, followed by an equally sudden decrease at 18,000 feet, the humidity subsequently increasing. The fluctuations in this series were numerous, there having been no fewer than four or perhaps five different strata of vapour."

It is evident that we have not at present sufficient data to draw safe conclusions as to the hygrometric condition of the superior atmospheric regions; and we trust that there will be continued series of observations undertaken, from time to time, under similar circumstances, and thus aërial voyages, hitherto barren of everything but risk, will be turned to good scientific account.

115. General remarks.—It may here be stated, that doubts have lately been thrown upon the accuracy of the dew-point temperature as deduced from the dry- and wet-bulb thermometer readings. M. Regnault, a great authority in these matters, considers that the wet-bulb temperature may not indicate accurately that of the air immediately surrounding it, and his condensing-hygrometer was the result of an attempt to produce an instrument which should give the dew-point directly, and not be liable to the objections brought against Daniell's, which we have already mentioned. It would appear, that at high elevations in the balloon ascent on Aug. 26th, whilst the indications of Regnault's hygrometer did not differ much from those of the dry- and wet-bulb thermometers at the height of 11,000 feet, the difference became considerable at about 12,000 or 13,000 feet, thus rendering it probable that at the latter heights the relative humidity, as deduced from the dry- and wet-bulb thermometers, was too great. The general accordance, however, was restored at 15,000 feet.

Col. Sykes, in his "Discussion of Meteorological Observations taken in India" (Phil. Trans. Part II. 1850), has entered upon the question at some length. He found that, in India, a very high degree of humidity resulted from observations with the dry- and wet-bulb thermometers, such as experience showed to be impossible; "and I have no hesitation," says he, "in expressing my belief that the results which I have obtained with the labour of some months (by reducing numerous observations with the dry- and wet-bulb thermometers) do not represent the real fractions of saturation of the air at the several places where the wet-bulb was observed."

As, however, in this country, we have a comparatively small range of temperature, extremes either of great heat or great cold being seldom reached, and as all the stations in the United Kingdom are at low elevations, we apprehend the majority of observers will follow the plan of observation with the dry-bulb and wet-bulb thermometers advantageously, both for the sake of uniformity, and from the small amount of expense and trouble which it involves. Nor are we without patronage as regards its adoption, seeing that at the late "Meteorological Congress" at Brussels, which was attended by men of eminence from the maritime nations of Europe and from the United States, the dry- and wet-bulb thermometers were recommended to be regularly observed; the chairman of the Congress, M. Quetelet, had long before thus expressed his opinion on the point now under discussion in his "Instructions pour l'observation des phénomènes périodiques," issued by the Académie Royale de Bruxelles:—

"L'hygromètre donne des renseignements

utiles; mais on le remplacera avantageusement par le psychromètre, moins sujet à se déranger, et dont les indications sont plus sûres."

No doubt can exist in the mind of any one who has examined and tested the Greenwich observations by Apjohn's formulæ, and who has compared them with the results arising from the use of Glaisher's Tables, that the subject of the hygrometrical state of the air is one which still demands the attention of physicists; the discrepancies in the results of various formulæ show clearly that, with regard to the deductions explained in this work, the most we can at present expect are approximations, and those by no means of the closest kind.

CLOUDS.

116. It may be fairly assumed that observations on the various kinds of clouds have not been followed up by scientific men with that attention which their evident connexion with atmospheric changes demands. Indeed the philosopher must acknowledge that the sailor and the fisherman on the coast are far more weatherwise than himself, and the clouds are that page of nature's book which only they are competent to read; dependent as they are on observation of the face of the heavens, they have attained by practice an aptitude in predicting atmospheric changes which puts science to the blush, and if we aim at ever arriving at the power of prediction as regards meteorological phenomena, we must mark and record more extensively the changes in the character of the clouds which float above us.

117. **Their appearance.**—The clouds are spread over the sky in forms of great variety, and, at times, of surpassing grandeur. Their beauty of colour, and the grand effects of contrast, defy imitation by the painter's art; they move above us mysteriously floating without visible support, and setting at naught the explanations which philosophers have given. We are told by some that they are vesicular vapour; this is simply a hypothesis; we can only affirm with certainty that they consist of particles of aqueous vapour in a peculiar state of aggregation, and that they float in the lower regions of the atmosphere. That electricity affects their state is pretty certain, but facts are wanting on which to found a theory as to its mode of operation.

118. **Height.**—The clouds do not seem to ascend higher in the atmosphere than five miles. Riccioli, who determined the height of clouds trigonometrically, never found them to reach higher than 8880 yards. Dalton found cirri, the lightest form of cloud, from three to five miles above the earth; and all observations combine to

fix the region of clouds in that stratum of the air included between the sea-level and five miles above it.

The existence of clouds does not seem to affect the atmospheric pressure; the barometer may remain stationary at times when heavy clouds are rolling above.

119. Classification.—In the year 1802, Mr. Luke Howard published an elaborate Essay "On the modifications of Clouds, and on the principles of their production, suspension, and destruction." The classification proposed by him has been generally adopted both in England and abroad; the following is a summary of his views, and an explanation of his nomenclature, which has the merit of being founded on the natural characteristics exhibited by clouds in the forms they assume, and on the causes from which they derive their origin.

The simple modifications are thus named and defined:—

1. *Cirrus.*—Parallel, flexuous, or diverging streaks or fibres of cloud.

2. *Cumulus.*—Convex or conical heaps, increasing upward from a horizontal base.

3. *Stratus.*—A widely extended, continuous, horizontal sheet.

The intermediate modifications are:—

4. *Cirro-cumulus.*—Small roundish masses of fleecy cloud, in close horizontal arrangement or contact.

5. *Cirro-stratus.*—Horizontal or slightly inclined masses attenuated towards a part or the whole of their circumference, or else a thin diffused sheet.

The compound modifications are:—

6. *Cumulo-stratus.*—The cirro-stratus blended with the cumulus, and either appearing intermixed with heaps of the latter, or superadding a wide-spread structure to it.

7. *Cumulo-cirro-stratus or Nimbus—the Rain-cloud.*—A cloud, or system of clouds, from which rain is falling; it is a horizontal sheet, above which spreads the cirrus.

120. *a.* **The Cirrus.**—This is the lightest of all clouds, and generally occupies the highest regions of that stratum of the air which alone is frequented by cloud; accurate measurements have proved this fact, which is confirmed by the circumstance of cirri reflecting the sun's rays for a long time after they have ceased to illumine the clouds below. Cirri have been known to present the same configuration for two successive days, while a strong breeze has been agitating the air beneath; hence it is probable that the vapour of which they are composed aggregates in

a calm region, perhaps out of the reach of the daily variations of temperature and evaporation which disturb the lower strata of the atmosphere. It may be that the electric states of dry air and of moist are antagonistic; the various fine ramifications may be the means of electric communication between the moist air of the cirrus and the dry air which surrounds it.

121. *b.* **The Cumulus.**—We have already seen that the sun's rays in traversing the atmosphere communicate little, if any, of their heat; it is by conduction from the earth's surface that the lower stratum of the air is heated, and the heat spreads by convection.

When the sun has risen, the surface of the earth becoming gradually warmed communicates heat to the stratum of air in contact with it, which, consequently, is capable of holding in suspension an increased amount of aqueous vapour; the current caused by the ascent of the heated air urges upwards the vapour, which during the night had remained in the air, till it arrives at a region so cold that it becomes condensed in part, and descends in fine particles, which are reabsorbed before they reach the earth; the meeting of the descending globules of vapour and of the ascending current forms a cloud, which gradually increases in size from attracting the vapour in its neighbourhood. Were the supply of vapour to proceed from every quarter, the form of such cloud would probably be spherical; but as the portion directed towards the earth is in contact with air not saturated with moisture, or having none to spare, we find the cumulus attaining only the hemispherical form, which is the general characteristic of this modification of cloud.

The Cumulus is formed only in the day-time, as then only it is that the disturbance in the temperature of the air above described can occur; it vanishes towards the evening from the superior strata of the air having increased in temperature, while the temperature of the inferior is being lowered, so that there will cease to be an upward current.

122. *c.* **The Stratus.**—This form is a horizontal sheet of cloud which occurs mostly at night or in the evening when the air is calm; it is at this time that it is formed, by the layers of air next the earth getting cooled from its contact, when the earth has lost heat by radiation, to below the Dew-point; it must therefore increase by additions to its upper surface, as higher portions of the air begin to lose their heat in turn. The Stratus comprehends mists which ascend from valleys, the surface of lakes and pieces of water, all which are dispersed by the heat of the morning sun.

123. Registration of amount of cloud.—In a register of meteorological phenomena, it is usual to place upon record the kind of cloud, if any, visible at the times of observation, and also the amount of cloud which overspreads the sky. A clear sky is registered 0, and a cloudy sky 10; and the observer, in intermediate conditions, estimates the amount of cloud and registers accordingly. From the Greenwich observations it has been concluded that a certain degree of regularity may be traced in the cloudiness of the sky, and from them has been formed a table of the "diurnal variation" in the amount of cloud, similar to that of the "diurnal range" of temperature. From this table it would appear that the hours of night are those in which the least amount of cloud prevails, and that the greatest amount may be expected at mid-day. The average amount of sky covered by cloud in the several months of the year is as follows:—

January	7·6	July...........	6·7
February......	7·5	August	6·4
March	6·4	September ...	5·9
April	5·9	October	6·8
May............	6·5	November ...	7·3
June	5·9	December ...	7·4

Hence it would appear that, from November to

February, three-fourths of the entire sky are covered by sun-repelling clouds.

The difference between the day and night, as regards the clearness of the sky, is exceedingly small during these months.

June is the clearest month in the year; during the night hours the least amount of cloud may be expected in summer and autumn, especially in the month of September, which is a month, as astronomers can testify, exceedingly favourable for observation.

Observers usually trust to estimation in recording the amount of cloud, which of course will only give a rough approximation. The author has been shown an ingenious contrivance by J. Campbell, Esq., of the Board of Health, by which the amount of sunshine has been made to register itself during the hours of day-light, well-worthy of mention in this place. A hollow globe of glass, 3 inches in diameter, is filled with Canada balsam (the aperture being covered with a piece of bladder) and exposed all day in the open air*; whatever may be the sun's declination, his rays will be concentrated in the focus of this spherical lens, which in this case will be distant from the

* The liquid need not be Canada balsam; water, with a little acid in it to keep it pure, will do.

surface of the sphere 0·615 in.*. The globe is fixed in a hollow hemisphere whose radius is greater than that of the globe by this quantity; a piece of black ribbon is laid on this opposite to the sun's diurnal arc for the day, and whenever the sun shines, his rays will be brought to a focus on it, and his progress will be marked by a line burnt through the ribbon if his light has been continuous, or by a broken line, or series of small holes, if it has been intermittent; a comparison of the length of these with the length of ribbon due to the number of hours the sun is above the horizon for the day, will give the amount of sunshine; and, as each strip of ribbon may be preserved, not only the number of hours of sunshine, but the exact time of their occurrence may be faithfully registered.

Mr. Campbell has given to the Royal Meteorological Society details of the construction and mode of use of this instrument, in a paper which is printed with the Report of the Council for 1857.

Foreign meteorologists register the "sérénité

* To find the focus of a sphere for parallel rays in terms of the radius, divide the index of refraction by twice its excess above 1. Refractive index of Canada balsam = 1.549.

$\therefore \frac{1{\cdot}549}{1{\cdot}098} \times 1{\cdot}5 = 2{\cdot}116$ inches, measured from the centre of the sphere;

$2{\cdot}116 - 1{\cdot}5 = 0{\cdot}616$ beyond the surface.

du ciel" or "portion of sky clear," and reverse the record used in England by letting 0 stand for a cloudy sky, and 10 for one of pure sunshine; this form was adopted by the Meteorological Congress at Brussels, in the directions which they published for the use of the naval officers of the whole civilized world. (See Part III.)

RAIN.—SNOW.—HAIL.

124. One of the most important elements in determining climate is a knowledge of the amount of rain received in any district;—to the engineer this is especially useful, as supplying data on which he may safely construct his sewers, or calculate the amount of water which will be afforded to a town within a given area. Of all meteorological observations, the determination of the quantity of rain which has fallen at any place within a given time is the easiest; of all instruments, the rain-gauge is the least expensive and least liable to be out of order.

125. Rain-gauge.—The principle of the instrument is the following:—If we imagine the surface of the ground, over which a shower of rain has passed, to be perfectly level and impervious to moisture, and that it is so surrounded with an enclosure that the whole quantity of water shall be retained, the rain would cover the surface to a certain depth, which, measured in inches, would give the amount of rain that had fallen. In calculating this depth by means of the rain-gauge, we expose a small surface to the reception of the rain, and measure the depth of what it receives, proceeding on the supposition that the same amount would have fallen into the gauge at any portion of the rain-fall; this equable distribution of rain, however, seldom occurs, for a shower may pass by the position chosen, and, although much rain may fall at no great distance, not a drop may reach the rain-gauge itself. Hence, to obtain the exact amount of rain which falls in any given district, several rain-gauges should be dispersed in various parts, and the mean of the whole amount received would be the true quantity due to such an area.

Observers are generally satisfied with registering the amount of rain received by their own gauges at 9 A.M. every day. In engineering operations, however, it is of the utmost importance to know how much rain may be expected on any occasion to fall per hour; hence, in very violent falls, the quantity should be ascertained immediately on their cessation, and the time of duration noted in the register.

Rain-gauges are of various constructions. In some a glass tube, divided into inches and parts of an inch, proceeds externally from the bottom of the vessel in which the rain is received, and is read off; after registration, the water is discharged by a stop-cock. The objection to this form of construction is the exposure to breakage of the glass tube on the occurrence of frost.

In others a float is elevated by the water, and the scale which is attached to it shows the depth of rain received.

Perhaps the most simple is the one which I have adopted; it is guarded from evaporation, when made of tin is exceedingly inexpensive, and is never liable to be out of order (Plate I. fig. 6.). A circular copper funnel, *a*, 12 inches in diameter, is connected by a pipe with a vessel, *b*, capable of holding a gallon or more. To the bottom of this vessel is attached a stop-cock, *c*, by means of which the rain is drawn off and measured in a graduated glass cylindrical jar, *e d* (fig. 7.). The divisions of the jar may be thus obtained: if *a* represent the diameter of the receiving vessel, and *b* that of the jar, *c* the depth of rain in the vessel, considered cylindrical, and *x* the required depth of the glass jar to measure such amount, then, since area, multiplied by the depth, gives the volume—

$$\cdot 7854a^2c = \cdot 7854b^2x; \quad \text{or, } a^2c = b^2x; \quad \text{or, } x = \frac{a^2}{b^2}c.$$

Now, suppose the diameter of the glass jar to

be 2 inches, and it is required to find what depth of the jar will measure $\frac{1}{4}$ of an inch, we have—

$$\frac{12^2 \times \cdot 25}{2^2} = x; \qquad \text{or, } x = 9.$$

Nine inches of the jar, 2 inches in diameter, will therefore measure one quarter of an inch of rain, received by a surface 12 inches in diameter. One twenty-fifth part of nine inches will consequently measure one-hundredth of an inch; and the thousandths may be estimated.

Among the contrivances—which have been numerous—for registering the amount of rain received and the time during which it falls, the most complete is "Osler's Pluviometer," which will be described in Part III., among the Instruments in use at the Royal Observatory, Greenwich.

Crosley's, also in use at the Observatory, is a self-registering rain-gauge. The collected water falls into a vibrating bucket; as soon as one side is full, the bucket oversets and presents another compartment, which, having received its portion, discharges it in an opposite direction. The bucket is thus, during the fall of rain, kept in a state of vibration. An anchor with pallets is attached to the axis on which it turns, which acts upon a toothed wheel by a process exactly the reverse of that of a clock-escapement. This wheel communicates motion to a train of wheels, each of which carries a hand upon a dial-plate, and thus inches, tenths, and hundredths are registered.

126. Position of rain-gauge.—The rain-gauge, for general registration, should be only a few feet from the ground, and in every case its height should be stated, as it is invariably found that more rain is received near the surface than at a superior elevation. Indeed, it should be agreed upon by observers that their gauges should all be at the same height, and all equally free from the interference of buildings or trees. Till some rule of this kind is adopted, we are not in a position to compare, so accurately as we might, the quantity of rain which falls in different districts. At Greenwich there are several rain-gauges at different heights above the ground.

The following table will show the differences between the quantities of rain received by them during 1846 and 1847:—

Height above the ground. ft.	in.	Inches of rain received in 1846.	Inches of rain received in 1847.
50	0	13·46	7·12
24	0	22·63	13·02
1	11	25·86	16·49
0	5½	25·29	17·61

Some observers prefer a small receiver, and consider that the amount of rain may be as readily decided with it as with a larger one. The great point to be attended to in the construction of receivers is that the area of the aperture for the reception of the rain be very carefully ascertained; the circular form is the best, as it admits of the rim being carefully turned in a lathe to the required dimensions; the rim should project over the body of the funnel to retain the water which splashes up, especially when the shower falls obliquely, and the divisions of the measuring-jar should be accurately tested by weighing the water received; indeed some observers always estimate the amount by weight, and the following considerations will show that this may be done with great accuracy and facility.

127. **Weight of rain-water.**—A cubic inch of distilled water at a temperature of 62° Fahr., according to the enactment of the British legislature, is a standard of weight; this quantity has been determined to weigh 252·458 grains, of which 437½ make one ounce avoirdupois; hence the volume of an ounce of water will be represented by $\frac{437\cdot5}{252\cdot458}$ inches. Let $w =$ the number of ounces in a quantity of water which we wish to measure; then, $\frac{437\cdot5}{252\cdot458}w$, will equal its volume.

Let a represent the area of a cylindrical vessel,

placed horizontally, and d the depth of water in it, then the volume of this quantity $= da$; consequently,

$$da = \frac{437{\cdot}5}{252{\cdot}458} \times w$$

is true for any quantity whatever. Putting r for the radius of the receiving vessel, supposed circular, and π for 3·1416, we have

$$d = \frac{437{\cdot}5}{252{\cdot}458} \times \frac{w}{a} = \frac{437{\cdot}5}{252{\cdot}458} \times \frac{w}{r^2 \pi}.$$

The only variable quantity in this expression is w, which may be obtained by weighing the amount of rain received; by combining the others, substituting for r the radius of any vessel we are making use of, we obtain a constant factor, which multiplied into w will give d.

128. Example.—At the Ordnance Map Office, Southampton, the rain is estimated by weight; the radius of the receiver is 2·75 inches, which, substituted for r in the formula, gives the factor ·07294. On the 5th of February, 1854, 2·63 ounces by weight fell on the receiver: then ·07294 × 2·63 = 0·192 inch, depth of rain fallen.

For the same shower the rain-gauge at my observatory * gave as the amount, 0·195.

* For this comparison I am indebted to Sergeant-Major Steel, of the Royal Sappers and Miners.

As rain-gauges are usually of 6, 8, or 12 inches diameter, the following factors have been computed for such dimensions.

129. Rule.—Weigh the rain received, in an accurate balance; multiply the weight in ounces avoirdupois by the factor determined for the diameter of the receiver, and the result will be the number of inches received.

Diameter 6 inches; factor ·06129.
Diameter 8 inches; factor ·03448.
* Diameter 12 inches; factor ·01532.

130. Relative amount of rain.—In the Reports of the British Association for 1851 and 1854 will be found contributions by the author towards the comparison of the climates of different places in England; those selected being Southampton, in the centre of the southern line of coast; Falmouth, on the extreme south-west; Stone, between Oxford and Aylesbury, a central situation; and York, a northern position, also inland. The fall of rain for five years is here given from those papers, together with the corresponding quantities at Greenwich; the columns headed "Days" denote the number of days in the year on which rain fell; those headed "Amount" the quantity of rain in inches.

* 5 in. factor ·08826

	1849.		1850.		1851.		1852.		1853.		Total.	
	Days.	Amount.	Days.	Amount.	Days.	Amount.	Days.	Amount.	Days.	Amount.	Days.	Amount.
Greenwich ..	153	23·8	141	19·7	146	20·5	153	34·4	184	29·0	777	127·4
Southampton	139	33	128	32·3	140	24·5	166	49·7	146	29·2	719	168·7
Falmouth....	191	38·6	163	38·7	184	37·2	191	50.	203	39·1	932	203·6
Stone	140	19·3	171	19·2	176	22·6	180	34·1	216	27·8	883	123
York........	160	23·5	159	17·9	136	20·5	156	27·2	174	22·3	785	111·4

The number of days on which 1 inch or more, and on which 0·5 inch or more, of rain fell within twenty-four hours, during these years, were at

	1849.		1850.		1851.		1852.		1853.		Total.	
	1 in.	0·5 in.	1 in.	0·5 in.	1 in.	0·5 in.	1 in.	0·5 in.	1 in.	0·5 in.	1 in.	0·5 in.
Greenwich ..	1	12	0	7	3	6	4	16	4	11	12	52
Southampton	2	21	4	20	3	9	8	39	2	21	19	110
Falmouth....	2	23	5	18	2	14	10	32	3	19	22	106
Stone	1	4	0	5	..	7	..	19	..	13	1	48
York........	0	5	0	0	0	7	2	9	0	5	2	26

The greatest amounts of rain known to have fallen in twenty-four hours during the five years were: at Greenwich, 2·63 inches; Southampton, 2·1 in.; Falmouth, 1·96 in.; Stone, 1·3 in.; York, 1·99 in.

131. Deductions.—From the above Tables we may deduce a few valuable results.

α. With regard to the number of falls of rain beyond half an inch in twenty-four hours, Southampton and Falmouth are about equal; the number of such days at Greenwich and Stone is about half of those at the former places; at York, one quarter.

β. The entire quantity of rain, during the five years, is greatest at Falmouth, 203·6 inches, which is 25 per cent. more than at Southampton, 66 per cent. above Greenwich and Stone, and nearly double the amount at York.

γ. The least number of rainy days occurs at Southampton; both at that place and Falmouth the rain falls in much larger quantities than at either of the others, which is explained by their proximity to the sea, and confirmed by the Table showing the number of days on which quantities exceeding one inch, and half an inch, were received.

On an average of thirty years it has been found that the mean annual fall of rain at Paris is 26·6 inches—the yearly amounts varying from 16·9 to 27·9 inches.

In the torrid zone rain falls in the greatest abundance; in parts of India 120 inches is an average yearly fall; in the year 1852, at Mahabaleshwar, 290 inches were registered, and it is said that on the hills north of Calcutta the enormous quantity of 600 inches had been reached in one year, 25 inches having fallen in a day.

132. Snow.—We are not at present able to say whether the flakes of snow are formed at once from the congelation of vapour in the cloud whence they fall, or whether the particles of vapour, at first minute, unite with other frozen particles as they pass through the successive strata of air, and thus gradually increase in magnitude. In the winter of 1853–4, some observations were made on the form of the snow-flakes, which were found more perfect at that season than ever before known; some photographic drawings of these flakes were distributed among the Members of the British Meteorological Society, for the purpose of directing their attention to the subject. On the morning of January 21, 1855, the thermometer being at 19°, a slight fall of snow occurred in a perfect calm; the flakes on this occasion were examined by the author with a microscope, and they presented most of those beautifully regular hexagonal forms of crystallization, which have been hitherto remarked only in the Arctic regions.

Snow is considered to yield $\frac{1}{10}$th of an inch of water for every inch in depth: thus, if the snow when melted and measured yields 1 inch of water, it is concluded that the fall was 10 inches in depth.

133. **Hail.**—The explanations which have been given of the phenomenon of hail are very unsatisfactory; that electricity is concerned in its formation is beyond a doubt. The melted hail is estimated in the rain-gauge as so much rain. Observers should be careful to record the phenomena attendant on hail-storms, such as the size and weight of the hail-stones, &c.; these at times have been known to weigh more than half a pound. Dr. Nöggerath, at Bonn, found the weight of hail-stones which fell on the 22nd of May, 1822, to be no less than 12 and 13 ounces.

3. *The Barometric Condition of the Air.*

134. **Fluctuations of the barometer.**—To the indications of the barometer alone are we indebted for any knowledge of what is taking place in the regions of the atmosphere far above us; the thermometer and hygrometer give us local determinations of the heat and humidity of that stratum of air which is in proximity to the ground, but,

as Humboldt observes, "Important changes of weather do not usually arise from a local cause situated at the place of observation itself; their origin is to be looked for in a disturbance of the equilibrium of the currents of the atmosphere, which has begun afar off, and generally not at the surface of the earth, but in the higher regions*." The barometer marks the heavings and pulsations of the atmosphere;—the preponderance or deficiency of air vertical to the place of observation;—the changes from a less pressure to a greater, and the reverse. We may regard the object of deductions from barometric observations to be twofold; the determination of the transmission of extensive atmospheric waves, and the registration of local variations in the atmospheric pressure, which occur regularly in successive periods of no great length.

135. Atmospheric waves.—The former branch of the subject has engaged the attention of Mr. Birt, who has published the result of his investigations from time to time in the "Reports" of the British Association. The method of tracing an atmospheric wave is by comparing the barometric registers kept at different points on the earth's surface, and observing the time when some remarkable rise and depression may have been re-

* 'Cosmos,' Sabine's translation.

corded in immediate succession at several places; there will be frequently found, when the diurnal and local interferences are eliminated, a wonderful agreement in the amount of such variation in pressure in situations far removed from each other; the depressions, or "troughs" of the wave, will be separated from the greatest pressure, or "crest" of the wave, by the same intervals of time, and the same, or nearly the same, amount of barometric pressure, at stations separated from each other by distances more or less considerable;—such waves have been traced from the western shore of the Atlantic to the eastern, and their identity proved.

136. Diurnal variation of the barometer.—Atmospheric waves of smaller size, and following a law dependent upon the sun's hour-angle, have been proved, by well-sustained observations, to be of a definite and regular character.

That a tidal wave of air, as well as water, should follow the moon in her daily course, was long suspected; it appears to have been proved by the observations at St. Helena since 1842, that the attraction of the moon causes the mercury in the barometer to stand, on the average, ·004 inch higher when the moon is on the meridian above or below the pole, than when she is six hours distant from the meridian. The small portion of this effect which would reach our latitude, would seem to be masked, at Greenwich, by the hourly fluctuations of the barometer, which are far less steady than within the tropics. The recorded diurnal oscillations of the barometer at Greenwich have been the foundation of a table of corrections to be applied to observations taken at any hour of local time, to obtain the mean height of the barometer for the month; in the same manner as the corrections for temperature explained in § 47.

The corrections for the hours of 3 A.M., 9 A.M., 3 P.M., and 9 P.M., are here given.

Local mean time.	Jan.	Feb.	March.	April.	May.	June.
h	in.	in.	in.	in.	in.	in.
3 A.M.	+·005	+·012	+·023	+·010	+·005	+·004
9	−·008	−·008	−·010	−·011	−·007	−·012
3 P.M.	+·004	+·006	+·003	+·009	+·006	+·007
9	−·007	−·008	−·015	−·009	−·006	+·003

Local mean time.	July.	August.	Sept.	October.	Nov.	Dec.
h	in.	in.	in.	in.	in.	in.
3 A.M.	+·005	+·011	+·010	+·015	+·008	+·010
9	−·010	−·008	−·011	−·009	−·005	−·010
3 P.M.	+·005	+·005	+·008	+·005	+·010	+·010
9	−·001	−·010	−·009	−·014	−·017	−·009

On frequent occasions I have applied these corrections separately to the monthly means of my 9 A.M., 3 P.M., and 9 P.M. observations, taken at Southampton, and the results were in no case consistent; nor is this surprising, when we regard the situation of Southampton at the head of an estuary which is divided into two arms by the Isle of Wight. Most probably the local variations of atmospheric pressure are peculiar, but what these may be can only be determined by a far more extensive series of observations than I have had the leisure to undertake.

137. Maxima and Minima.—At Greenwich, the barometer readings show a double maximum and minimum in the twenty-four hours; unlike the curve of mean temperature, projected in Plate IV. fig. 2, the daily average readings of the barometer projected for any month, would form a curve with two ascending branches and two descending—with two apices, and two corresponding depressions, while four times a day the reading of the barometer would be at its mean value; one of these mean readings will occur with the greatest steadiness some time between mid-day and 2 P.M., according to the season.

The horary variations of the barometer within the tropics are exceedingly regular; they present two maxima, at 9 A.M. and 10½ P.M.; and two minima, at 4 A.M. and 4 P.M., which latter are nearly the hottest and coldest hours of the day.

138. Explanation.—At stations in the interior of great continents far removed from the sea, the double maximum and minimum disappears, and the diurnal variation exhibits a single maximum and minimum, similar to that of the diurnal curve of mean temperature; the turning-points of the barometric curve seem nearly to correspond with the times of greatest and least heat. The explanation of this phenomenon would seem to be, that the air superincumbent on that portion of the earth which is heated by the sun, rises, and extending in height overflows laterally, causing additional pressure all around, but a diminished pressure at the base of the column, which will reach its minimum at a period of the day not far removed from that of the maximum of heat. As the surface cools in the latter part of the day, the air above it cools also and descends, receiving the overflow from the adjacent air, which has become heated in its turn on the rotation of the earth; and thus the pressure will gradually increase, and arrive at its maximum at a period of the twenty-four hours not far removed from the minimum of heat.

It is presumed that the double barometric maxima and minima, observed at stations in the neighbourhood of an abundant supply of water for evaporation, may be explained by separating the pressure of air from that of water, when it would be found that one of the maximum and one of the minimum points would be due to the rise and fall of the gaseous atmosphere, and the others to the variations in the amount of aqueous vapour mixed with it. Comparisons have been instituted between the amount of gaseous and vapour pressure, and the projections of the curve of these pressures separately, combined with the direction of the wind, have tended to give a confidence in this explanation, which will be increased by inspecting the curves projected by Col. Sabine, in the British Association's Report for 1845, from the data supplied by the meteorological observations at Bombay.

The same principles would appear to account for the annual variations in the barometer, the monthly means being dependent on the amount of aqueous vapour, combined with the gaseous atmosphere.

139. **Character of instruments.**—As the object of this work, however, is rather to show how to observe phenomena, than to enter at large upon their explanation, we must proceed to explain the instruments of observation; the very subjects we have been discussing, viz. the turning-points of the barometric curve, will show how necessary it is that such instruments should be capable of very nice indications, for at Greenwich the difference between the highest and lowest reading of the

barometer will, in some months, not amount to 0·02 inch in the twenty-four hours, as far as diurnal variation is concerned. With regard to the times of maximum and minimum pressure, much would seem to depend on the sensitiveness of the instrument. The water-barometer, constructed by the late Professor Daniell, arrived at its maximum an hour before the standard of the Royal Society, while this preceded a mountain barometer by the same interval: had a marine barometer been remarked at the same time, it would probably have been still further behind, as it is, from its construction, the least sensitive of any. We see, then, the importance, in all barometric registers, of inserting a full and complete description of the barometer employed, the size of its tube, the mode of measurement by the scale, the maker's name, and the fact whether it has or not been compared with some well-known and acknowledged standard, such as the flint-glass barometer of the Royal Society: we then have the means of judging of the worth of a series of observations, which no labour can render valuable if the instrument be not one whose character will stand a severe test.

140. Corrections of the barometric reading.—In fig. 1, Plate I. we have a representation of the barometer in its simplest form; after the lapse of 200 years, the standard barometer, the tube of which is 1 inch in diameter, just erected at Kew under the sanction of the most eminent physicists, is an exact reproduction of the original experiment by which Torricelli convinced himself of the weight of the air. All the varieties of construction have aimed only at ensuring, in the most accurate way, a correct measurement of the height of the column *c d*; in arriving at this determination with one reading of any scale, we encounter difficulties and incur errors, the nature of which must be detected and their value ascertained.

The vacuum at the upper part of the tube, if the instrument is well constructed, though the most perfect that can be produced, is not complete, for the space above the column is filled with the vapour of mercury, though of a very low tension dependent upon the temperature; to prevent the rise of particles of air which may be diffused throughout the mercury, or may have been attached to the sides of the tube, the mercury should be boiled in the tube, and the perfection of the vacuum may be tested by inclining the tube and driving the mercury to the closed end, on striking which it will give a sharp and sudden tap if no air or moisture exist above the mercurial column.

141. Capillarity.—If a piece of glass tube, not more, we will suppose, than ·4 inch in diameter, be inserted in water, the water will rise within it by capillary attraction to a height greater or less according to the size of the tube, the surface of the water within being concave; on the contrary, if the same tube be plunged into mercury, it will repel the metal all around, and the surface of the mercury within the tube will be convex, the top of the curve being depressed below the level of the liquid in the vessel. Now, unless the tube of the barometer is so large* that the capillary action may be disregarded, it is evident that a correction must be applied to the observed height of the mercury in the barometer to reduce it to the true. This correction is always +, and is usually determined by the maker; if it be not, it may readily be obtained from tables when the diameter of the tube is known.

142. Temperature.—The scales of barometers adapted to scientific use are of brass throughout, extending from the cistern to the top of the tube; an increase of heat will be followed by an expansion both of the mercury and the scale. If the two metals expanded equally for equal increments of heat, no error would arise; but mercury expands more than any other metal known. Now supposing the atmospheric pressure to remain the same, but that the temperature has risen within

* The diameter of the tube of the Greenwich standard is 0·565 inch; the correction for capillarity is 0·002 inch.

a given period from 40° to 60°, the index would show (at a height of about 30 inches) a rise of 0·054 inch, which would be due, not to increased pressure, but to the excess of the expansion of the mercury over that of the brass scale. It has been agreed to reduce all observations to a standard temperature, viz. 32° Fahrenheit—the freezing-point of water—and for this purpose corrections are given, as in Table VII. (Appendix), which may be obtained by inspection.

In most barometers a thermometer with its bulb in the cistern shows the temperature of the mercury, and it is presumed that this is the same throughout the column. Sir John Herschel* objects to this arrangement, on the ground that the thermometer does not give the temperature of the whole mass including the column. In barometers lately constructed by Negretti and Zambra, the thermometer bulb is of the same diameter as the barometer tube, and cased in brass, so as to be, as much as possible, similarly circumstanced; and it is presumed that, unless in cases of sudden changes, the temperature of the two portions of mercury will be the same. It is advisable that the barometer be suspended in a room whose temperature is not liable to sudden variation, that error may not arise from this source.

* Admiralty Manual of Scientific Inquiry.

143. Capacity.—When the atmospheric pressure diminishes, the mercury sinks in the tube and rises in the cistern. The height measured by the scale, supposing it to be fixed, will not then be the true, as its divisions presume the level of the surface to be constant and not fluctuating—in fact, there will only be one point at which the measured distance will exactly agree with the real distance of the top of the column from the surface of the mercury in the cistern. This is termed the *neutral point*, and is ascertained experimentally by the maker during the progress of construction and engraved on the scale, together with the proportion between the area of a section of the tube and a section of the cistern. It is evident that the surface of the mercury in the cistern will be lower than the zero-point of the scale when the reading is above the neutral point, from the abstraction of a portion of its contents to supply the rise in the tube; and that it will be higher when the reading is below the neutral point. If the capacities be as one to forty-two, one forty-second part of the difference between the neutral point and any particular reading must be added in the former case, and subtracted in the latter, to obtain a corrected height.

To the barometer with which my observations at Southampton were taken, which was a very excellent one made specially for me by Newman, it was necessary to apply the capacity correction. I have tabulated the capacity and capillarity corrections and index error, so that the algebraic sum of a constant quantity and the temperature correction, applied to the reading of the vernier, will always give me the true height. In lieu of the glass cistern and leathern bag with which many barometers are supplied, a double iron cistern with a solid bottom is introduced, and, with great simplicity, the mercury is secured for travelling by stopping off the greater portion after the instrument is inverted. It has accompanied me several hundred miles, and, without requiring special care, has returned uninjured. I have applied it to the purpose of measuring heights with great success, and have the utmost confidence in its indications. During the months of January, February, and March, 1854, simultaneous readings were taken with it at my observatory at 9 A.M., and with one of Barrow's (described hereafter) at the Ordnance Map Office, Southampton: the differences between the mean readings for these months were (mine being in excess) ·024, ·030, ·025; the mean of these is ·026,— exactly the difference which, by calculation, is due to the difference of level between the two positions.

The marine barometers supplied to Government by Mr. Adie, are so divided that the inches are not really inches, but their representatives, each one being an inch minus the capacity correction.

The barometer has a portion of its tube contracted to a very fine bore, to prevent the oscillation, or "pumping," of the mercury from the motion of the vessel. In the barometer under consideration, which the Governments of England and the United States have adopted at the recommendation of the Kew Committee of the British Association, a pipette, or Gay-Lussac's air-trap, is inserted a little below the contraction, which prevents the entrance of air into the upper part of the tube; the tube is fitted into the cast-iron cylinder with cement, a portion of the upper part of the cistern being covered with strong sheep-skin leather, which will admit the air but not allow the mercury to pass, so that the instrument will sustain no injury by being laid horizontally; the attached thermometer bulb is included within the brass case, but does not dip into the cistern itself. All these as they are sent out will have been verified by comparison with the standard barometer at Kew. The appearance of the instrument, whose scale is of brass, is not unlike the usual mountain barometer.

The British Meteorological Society recommend

a barometer by Barrow, of Oxenden street—all those in use among the members having been compared either directly or intermediately with the Royal Society's Standard, and their index errors determined. The tube is enclosed in a hollow brass cylindrical case, on which is engraved the scale, which, as usual, reads to ·002 inch; the cistern consists of a hollow cylinder of glass closed by a leathern bottom. A small ivory index points downward towards the surface of the mercury, and the first step in taking an observation is, by means of a screw which acts on the leathern extremity of the cistern, to adjust the level of the mercury until it exactly touches the ivory point; the same action either raises or depresses the column of mercury, and, as the extremity of the ivory point is the zero of the scale, the reading will show the real height of the mercurial column above the surface of the liquid metal in the cistern, subject to only two corrections, viz. for temperature and capillary action. The barometer is attached to a mahogany slab, projecting forwards about 2 inches, and is free to turn on its axis in any direction. In reading off the scale a moveable ring is made to form a tangent to the curved surface of the mercury in the tube, a piece of white paper to reflect the light being placed behind it; with very little trouble the temperature and capillarity correction may be combined in one, together with a small zero correction to reduce the reading to the R. S. standard. When a table is thus formed, the absolute height of the barometric column may be ascertained from the reading by the application of one correction only, and that, it is presumed, with the accuracy which has hitherto been attained by standard barometers at three times the price.

144. Mountain Barometer.—The beautiful mountain barometers of Troughton and Sims are adjusted by means of a screw which urges the mercury upwards, so that it may fill the whole tube and render the instrument portable, and it also serves, at every observation, to adjust the surface of the mercury in the cistern to the zero of the scale; which, however, is not an ivory point. In the brass box which covers the glass cistern of mercury, near the bottom of the tube, are two slits made horizontally, precisely similar and opposite to each other, the plane of the upper edges of which represents the beginning of the scale of inches, or the zero of the barometer; before reading off the height of the column, the mercury is brought by the screw in such a position as exactly to shut out the light between its surface and the edges of the slits; the column then read off will give the true height of the mercury. The scale of the mountain barometer employed in determining heights, must have a greater range of inches engraved on it than is necessary for one used only for general purposes; an extent from 28 to 31 inches will embrace all the fluctuations to which the stratum of atmosphere, not far removed from the sea-level, will be subject.

145. Measurement of heights.—The measurement of heights by the barometer is a most useful application of the instrument, and various have been the formulæ proposed for the solution of the problem; they have been partly empirical, and partly dependent on the principle that, inasmuch as the heights increase in arithmetical as the densities diminish in geometrical proportion, if we suppose the densities to be a series of natural numbers, the heights corresponding to the densities will be the logarithms of those numbers. The mathematical reasoning on the subject must be omitted in this place*; and we shall give one formula only, selecting that of La Place, which was used by Mr. Welsh in calculating the heights reached in the balloon ascents of 1852. It has the advantage of being solvable by a table of common logarithms, and it takes in every minute consideration in the corrections which it demands.

* Narrien's 'Practical Astronomy and Geodesy' may be consulted on the subject.

Let z = the height required; h and h' the readings of the barometer corrected for temperature; t and t' the temperature of the air, at the lower and upper stations respectively; L the latitude; then

$$z = \frac{20939151n}{20886900 - n};$$

where the value of n is derived from the formula

$$\log\left(\frac{h}{h'}\right) \times 60159\left(1 + \frac{t + t' - 64}{900}\right)(1 + 0{\cdot}02837 \cos 2L).$$

An example worked out in full will show the application of the formula.

November 10th, 1852, on the earth's surface, 120 feet above the sea-level, at 2 hours 30 minutes 35 seconds P.M., the barometer reading was 29·978 inches $=h$; the air-temperature 49°·7 $=t$; at the elevation the balloon had reached at that instant the barometer stood at 23·45 $=h'$, the temperature being 35°·6 $=t'$; required the height of the balloon, in feet, above the station at which the corresponding observations were made :—

h log	1·4768027		
h' ar. co.	8·6298572		
	0·1066599	log	9·0280016
	60159	„	4·7793006
	1·0237	„	0·0101727
	1·000638	„	0·0002769
	n=6573	„	3·8177518
	20939151	„	7·3209559
	20880327	ar. co.	2·6802630
	z=6591	„	3·8189707

As it may frequently occur that an approximation to the true altitude of any position above another may be desirable without the trouble of a long computation, the author, after having given some considerable attention to the subject, has calculated the following Table, which he believes will be found correct for heights not exceeding 5000 feet, beyond which elevations in this country do not rise. The theorem accompanying it is from Sir George Shuckburgh, as is also the height of a column of air which shall cause a rise of 1 inch of mercury at a temperature of 32°, viz. 868·5 feet; the other quantities are tabulated on the assumption, founded on Regnault's determination, that air expands $\frac{1}{491 \cdot 2}$ of its bulk for every increase of 1° of temperature.

Let z, h, h', t and t' be as before, and T the tabular quantity opposite $\frac{t+t'}{2}$, or the mean of the temperatures of the air at the upper and lower stations; then

$$z = \frac{30(h-h')\mathrm{T}}{\frac{h+h'}{2}}.$$

Table showing the height, in feet, of a column of air equivalent in weight to a column of mercury 1 inch in height, at every degree of temperature from 30° to 81°, the barometric pressure being 30 inches.

Temp.	Feet.	Temp.	Feet.	Temp.	Feet.	Temp.	Feet.
30°	865·1	43°	888·0	56°	911·1	69°	934·1
31	866·8	44	889·8	57	912·9	70	935·8
32	868·5	45	891·6	58	914·7	71	937·5
33	870·3	46	893·4	59	916·5	72	939·3
34	872·1	47	895·2	60	918·2	73	941·1
35	873·9	48	897·0	61	919·9	74	942·9
36	875·7	49	898·8	62	921·6	75	944·7
37	877·5	50	900·5	63	923·4	76	946·5
38	879·3	51	902·2	64	925·2	77	948·3
39	881·1	52	903·9	65	927·0	78	950·1
40	882·8	53	905·7	66	928·8	79	951·8
41	884·5	54	907·5	67	930·6	80	953·5
42	886·2	55	909·3	68	932·4	81	955·2

To test this method I will select an example, which may be found worked out in Mr. Sims's treatise on mathematical instruments by Baily's formula.

The following observations were made in the Transit Room of the Royal Observatory, and at the base of the statue of George II., in Greenwich Hospital, to determine the difference of altitude.

	Upper Station.	Lower Station.
Detached therm. ...	$71°\cdot5=t'$	$71°\cdot5=t$
Attached ditto	70°	70°
Barometer	29·870 in.$=h'$	30·014 in.$=h$:

here $$z=\frac{30\times\cdot144\times938}{29\cdot942}=135\cdot4.$$

The difference of altitude as obtained by levelling with the spirit-level (Phil. Trans. 1831, Part I.) =135·57 feet.

The observations for determining heights with the barometer should in strictness be taken at the upper and lower stations simultaneously; for which purpose two observers and two instruments are necessary. When one person takes the observations, the barometer reading should be first recorded at the lower station, then at the higher, then without loss of time the lower reading should be noted on descending; should the barometer be steady, the mean of the two readings at the lower position will come very near the truth.

146. Reduction to the sea-level.—The Table, page 215, will serve to reduce barometer readings taken at any known elevation above the sea, to what they would have been if taken at the sea-level. Let T be the tabular number opposite the temperature of the air; h the reading of the barometer at f feet above the sea-level, and x the correction required;

then $$x=\frac{f}{T}\times\frac{h}{30}.$$

Suppose the barometer to read 29·500 at 60 feet above the sea, the air-temperature being 50°; what would be the reading at the sea-level?—

$$x=\frac{60}{900\cdot5}\times\frac{29\cdot5}{30}=\cdot065 \text{ correction required;}$$

therefore the reading at the sea-level will be 29·500+·065=29·565.

147. Adie's Sympiesometer.—As the transport of the barometer from place to place is attended with considerable trouble and demands great care, other instruments less liable to derangement have been contrived to express the pressure of the air. The sympiesometer of Mr. Adie of Edinburgh accomplishes this by allowing a portion of air or hydrogen gas to be compressed into a small chamber by the pressure of the atmosphere on the surface of a fluid, as shown in Plate II. fig. 4. ABDC is a glass tube from 9 to 18 inches in length, terminating upward in a bulb C, and, at the lower end, in the cistern AB; the tube is partly filled with oil, or coloured sulphuric acid, which, on being urged upward by the atmospheric pressure, compresses the air in the upper part of the tube into a space small in proportion as the pressure is great. To correct the error which would arise from the change of volume in the enclosed air produced by change of temperature, the scale EF, which carries an index, *a*, is moveable, and before reading off this index must be brought, on the scale GH, to correspond with the air-temperature, as shown by the thermometer IK. The index *c*, which is moveable on the tube, being then brought to the upper surface of the liquid column, will cut the scale EF at the reading due to the atmospheric pressure at the time of observation.

This instrument is much used at sea; not that it will compare in correctness with a barometer, but being easily affected by atmospheric changes it gives an early notice, and thereby directs attention to the marine barometer itself, the indications of which are sluggish, from its contracted bore.

148. The Aneroid Barometer.—This elegant instrument was invented a few years since by M. Vidi of Paris. In its latest form it consists of a cylindrical case about 4 inches in diameter and $1\frac{1}{4}$ deep, in which lies a thin metal box, near to and parallel with the curved boundary of the case, its two ends being distant about half an inch from each other. From this box the air has been exhausted, and the pressure of the external atmosphere on it causes it to alter its form, and to increase in length to a small extent, as the pressure becomes greater; by the intervention of a system of levers the small expansion is rendered very evident; the last of these is curved, toothed, and fits into a pinion concentric with the index, which turns with it and traverses a dial. The index must first be set to the correct reading by a standard barometer, for which provision is made in the construction; after this the instrument will give a very near approximation to the reading of a good barometer, nor will this reading be very much affected by the usual range of temperature. If the hand gets shifted by a shake, or in any other way, it can be re-set by a turn of the screw at the back of the instrument. Its portability renders the aneroid an agreeable companion for the tourist; its indications will show a rise of even a few feet, and hence, for general purposes, its utility in determining heights, when extreme accuracy is not required and the range is inconsiderable, is very great.

As no correction need be applied for temperature as regards instrumental variation, a simpler formula may be adopted than for a mercurial barometer; perhaps the best is one by Poisson, from which Prof. Patton (in the Journ. of the Geogr. Soc. vol. xxi.) has derived the following rule.

Multiply the number in the following Table opposite to the mean of the temperatures of the air at the two places (in degrees of Fahrenheit) by the difference of the barometric heights, and divide by their sum. The quotient will be the height in feet.

TABLE.

°		°		°	
32	52416	52	54745	72	57055
33	52532	53	54862	73	57192
34	52649	54	54979	74	57308
35	52765	55	55095	75	57424
36	52882	56	55211	76	57541
37	52993	57	55328	77	57658
38	53115	58	55444	78	57774
39	53231	59	55561	79	57890
40	53348	60	55677	80	58007
41	53464	61	55794	81	58124
42	53581	62	55901	82	58240
43	53697	63	56027	83	58356
44	53814	64	56143	84	58472
45	53930	65	56260	85	58539
46	54046	66	56376	86	58706
47	54163	67	56493	87	58823
48	54280	68	56609	88	58939
49	54396	69	56720	89	59055
50	54512	70	56842	90	59172
51	54629	71	56959	91	59288

This Table is applicable to observations with a mercurial barometer if they be first reduced to a standard temperature, which may be done by means of Table VII. (Appendix), as explained in p. 206.

149. Heights determined by the boiling-point of water.—When the tension of vapour of water is equal to the atmospheric pressure, ebullition takes place; it follows that if the pressure be reduced, water will boil at a lower temperature than that necessary to produce ebullition under the higher pressure. On this ground has been founded a method of determining heights by noting the temperature of the boiling-point at different elevations above the earth's surface. The formula of De Luc, reduced to English measures, is the one generally adopted. It proceeds on the assumption that the boiling-point will be reduced in temperature one degree for 548 feet of additional elevation; then letting H stand for the number of English feet in vertical height between two stations; b and b' the boiling-points at the lower and upper stations respectively; t the mean temperature of the air,

$$H=548(b-b')\left\{1+(t-32^\circ)\cdot 00222\right\}.$$

150. Example.—In the Phil. Trans. Part II. 1846, will be found a paper on the subject by Professor Christie, of the Royal Military Academy, from which is extracted the following example: in that paper the apparatus used is described at full, and the results of various measurements on the Swiss Alps recorded.

At Geneva the observed boiling-point of water was $209^\circ\cdot 335$; on the Great St. Bernard, $197^\circ\cdot 64$; the mean temperature of the air being $63^\circ\cdot 5$: required the height of the Great St. Bernard above Geneva. By substitution, we get

$$H=548\times 11\cdot 695\times 1\cdot 07=6857\cdot 5 \text{ feet.}$$

Prof. J. D. Forbes's observations on the Alps

have convinced him that, for heights not above 12,000 ft., accuracy enough can be got by simply multiplying 543·2 ft. by the number of degrees that the boiling-point is below 212°.

151. Barometric indications.—That change of weather is indicated by the barometer is undoubted; with our present knowledge of the laws of the atmosphere we are far from being able to predict with certainty from its indications. A few principles may be received as of general, though by no means universal, application; and they are here given as the result of the experience and observation of former ages; not that too much reliance ought to be placed on them, as no rule will hold good in every instance.

α. Changes of weather are indicated by changes in the height of the column, and not by its absolute height. However, when the mercury is low, wind and perhaps storms may be anticipated.

β. Generally the rising of the mercury indicates the approach of fair weather; the falling of it shows the approach of foul weather.

γ. In sultry weather the fall of the mercury indicates coming thunder. In winter the rise of the mercury indicates frost. In frost its fall indicates thaw, and its rise indicates snow.

δ. Whatever change of weather suddenly follows a change in the barometer, it may be expected to last but a short time. Thus, if fair weather follow immediately the rise of the mercury, there will be very little of it; and in the same way, if foul weather follow the fall of the mercury, it will last but a short time.

ε. If fair weather continue for several days during which the mercury continually falls, a long succession of foul weather will probably ensue; and again, if foul weather continue for several days, while the mercury continually rises, a long succession of fair weather will probably succeed.

ζ. A fluctuating and unsettled state in the mercurial column indicates changeable weather.

The following 'Card to accompany Weather-Glasses' is taken from the Board of Trade Meteorological Papers, in the use of which it must be remembered that the Barometer *foretells* weather rather than indicates its present state.

The Barometer rises for North-easterly wind (including from N.W., by the N. to the Eastward),—for dry or less wet weather,—for less wind,—or for more than one of these changes.	The Barometer falls for South-westerly wind (including from S.E., by the S. to the Westward),—for wet weather,—for stronger wind,—or for more than one of these changes.
Except on a few occasions, when rain (or snow) comes from the North-eastward with *strong* wind.	Except on a few occasions, when *moderate* wind with rain (or snow) comes from the North-eastward.

4. *Electric Condition of the Air.*

152. Nature of atmospheric electricity.—Electricity pervades the atmosphere at all times, and its presence may be detected with facility; its more stupendous phenomena, as exhibited in the thunder-storm, are not of frequent occurrence, but observation has shown that a certain electric action is constantly in progress, even when the sky is serene, though the source and the mode of operation of atmospheric electricity are most mysterious.

In experiments on atmospheric electricity, as usually conducted, a pointed metallic wire is elevated to some considerable height above the earth's surface; this wire serves to conduct the electricity within the reach of the observer, who may then apply his tests and his measurements for determining its nature and intensity. The phenomena developed by electricity thus collected from the air are precisely those of statical or frictional electricity, and the various experiments of the lecture-table may be reproduced by that collected from the air.

Experiments would seem to demonstrate that the electricity developed on the surface of the earth is negative, while the atmosphere is in general positively charged; the disturbance of the equilibrium between these two opposite states, gives rise to all the phenomena of atmospheric electricity. Sometimes, as in foggy weather, the nature of the electric charges will be changed, the earth becoming positive and the air negative. During a storm of rain, though the air may exhibit symptoms of being positively charged, the drops of rain, on examination, will be frequently found to be electrified negatively.

The clouds which float far above the earth's surface may be, and indeed often are, in different electrical conditions. Those similarly electrified will repel, those in opposite states will attract each other. If a stratum of dry air intervene between two clouds in opposite electric conditions, the equilibrium will be restored, when they are so near each other that the force, or intensity, of the electricity overcomes the resistance of the intervening air, by a violent discharge, producing the phenomena of thunder and lightning; as clouds, however, are very imperfect conductors of electricity, the equilibrium will only be established between certain portions of the opposing clouds, and can only be completed by successive discharges; and these are indicated by the repeated flashes and reports which constitute a thunderstorm.

If between two clouds in opposite electric states a stratum of moist air intervene, their conditions will be equalized quietly and insensibly by conduction; this would appear to be the method by which the change of condition in the lower strata of air, which is all we are at present prepared to record, is brought about, the tendency being to equalize the positive and negative electricity that prevails, the one in the air, the other on the surface of the earth.

Observers of the phenomena of atmospheric electricity have endeavoured to ascertain the nature and amount exhibited, the periods of the day and year when the intensity is greatest and least, and to trace, if possible, a connexion between the development of electricity and other atmospheric phenomena; especial care being taken to record changes in character and intensity during storms of rain, hail, thunder, and violent gales of wind.

153. Quetelet's deductions.—M. Quetelet has carried on a series of daily observations on atmospheric electricity at Brussels for ten years, and some of his conclusions confirm those to which Saussure arrived at an early period.

a. He concludes when the air is calm and clear, and free from the interference of high buildings, elevations, or trees, which might serve as conductors, it is always charged with electricity, which is for the most part positive.

β. That the electricity of the air attains two maxima of intensity—at a short period after sunrise and sunset—and two minima; the first occurring at about 2 or 3 P.M., and the second during the night. The times of maximum intensity observed at Kew are, in the summer, 10 A.M. and 10 P.M.; in the winter, 10 A.M. and 8 P.M.: the times of the minima are, in the summer, 2 A.M. and 10 P.M.; in the winter, 4 A.M. and 8 P.M.

γ. In the winter the atmospheric electricity exhibits greater intensity than in the summer; the intensity in January being to that of July in the proportion of 13 to 1.

Electrical Apparatus at the Kew Observatory.

154. The Observatory at Kew is a building the property of the Government, which has been lent to the British Association for the Advancement of Science for some years, for the purpose not so much of carrying on an uninterrupted series of observations on atmospheric phenomena, as for enabling the members to form a sound judgment on various instruments submitted to trial and comparison in that place. It was for many years under the superintendence of Mr. Ronalds (assisted by Mr. Birt), whose skill and ingenuity have displayed themselves in various contrivances connected with observation. It is not consistent with my design to notice all that this gentleman has done for science, but I cannot forbear alluding to a very neat contrivance by which the variations of the magnet are daguerreotyped on a plate of prepared metal, which is moved by clock-work and thus forms an accurate register of its oscillations. The whole apparatus is very compact, and the Observatory at Toronto has lately been supplied with one by the British Government. The account may be found in full in the Philosophical Transactions, Part I. 1847.

Mr. Ronalds has invented a self-adjusting barometer, in which the expansion of the mercury by heat is counteracted by bars of zinc, a metal which expands about one-sixth part of the expansion of mercury at the same temperature; as these dilate, they put in motion a system of compound levers, to which the barometer-tube is attached, and thus the surface of the mercury is always kept at the same distance from the zero of the scale as it would be were it not to be subject to expansion with an increase of temperature.

The branch of atmospheric investigation in which the merits of Mr. Ronalds are most extensively recognized, is electricity. In the year 1843, the Observatory was completely fitted with every requisite for judging of the electric state of the atmosphere, and the contrivances were found to answer their purpose so well, that those adopted at Greenwich, Toronto, and elsewhere, were copied from them, and constructed under Mr. Ronalds's superintendence. A description, therefore, of those in use at Kew will be all that can be required to give a general view of what is doing in the best observatories throughout the British Empire, in the registry of the electric state of the atmosphere.

Mr. Ronalds some years since completed a contrivance by which the state of the electrometer is registered by the daguerreotype. The prepared plate of metal is suspended vertically, and is drawn up by clock-work, and the gold-leaf electrometer is interposed between it and the light. On the opening of the leaves a mark is left on the surface of the metal plate, exactly at that spot which corresponds to the time when the occurrence took place and the duration of the electric action. This instrument is described in the paper just referred to. The electric apparatus now about to be described is marked by complete efficiency, as well as by the compactness and simplicity of all the arrangements. The dome in which it is located rises high above the rest of the observatory, and there are no buildings or trees to interfere with the full development of the electricity existing in the air. We may therefore conclude that the records deserve our full credence, both from the nature of the instruments and the character of the observers, and from the favourable situation in which the observations have been taken.

Mr. Ronalds's attention was for some time directed to experiments on "frequency" of atmospheric electricity, that is, the rate at which a new charge rises to its maximum after the former charge of an atmospheric insulated conductor has been destroyed. The observations were taken at such periods of the day as sunrise, noon, and sunset.

For the record of the rapidly succeeding and varied electric phenomena during the passage of a storm, he introduced what he terms a "storm-clock," without which it would be impossible for one observer to register the observations. It consists of a time-piece, which carries an index down a long sheet of paper laid on a desk; this it accomplishes in half an hour, and the observer has simply to record the events as rapidly as they occur opposite to the point of the index, which can evidently be done much more readily than by reading the chronometer and setting down the time at successive instants. In the hurry of the moment mistakes are often made, and several phenomena are entirely lost; whereas one observer, by means of this contrivance, accomplishes as much work as two could effect in the usual method.

Plate VII. represents the dome of the Observatory at Kew, with the electrical apparatus *in situ*; through the centre of the dome a circular aperture has been cut, in which is fitted a mahogany varnished cylinder, *a, a'*. G, G is a strong cylindrical pedestal, which serves as a closet for articles connected with the observations. It is surrounded by a stage, which, as well as the steps by which the observer ascends, is detached. C, C is the safety-conductor, for conveying the electricity away from the building. The principal conductor, D, D, is a conical tube of thin copper 16 feet high; E is a brass tube into which it is firmly secured; F is a hollow glass pillar, the lower end of which is trumpet-shaped and ground flat. A collar of thick leather is interposed between F and the table, and such is the firmness of the whole that the conductor has resisted gales which have uprooted trees in the neighbourhood. H is a spherical ring carrying four arms at right angles to each other, three of which are shown in the engraving; I, I are two of these. *k* is a lamp for warming the glass tube F, in order to produce perfect insulation; K is a chimney of copper, closed above, passing through the table and entering but not touching F. By this arrangement the lower part of F is generally warmed too much and the upper too little; but the pillar F, being conical, some zone always exists between the two ends, which is in the best possible state for electrical insulation. L is a set of finely-pointed platinum wires soldered to D. M is a Volta's small lantern. N is an inverted copper dish or parapluie, fitted by a collar and stays on E, and of course insulated by F; its least distance from the mahogany cylinder is 3 inches. It will be seen that, by this arrangement, the active parts of all the electrometers and the conductor itself are insulated by the glass pillar. O is a Volta's electrometer, No. 1; P, Volta's electrometer, No. 2; Q, Henley's electrometer; S, a galvanometer by M. Gougon. No. 1 is the most sensitive, and comes into action first; No. 2 then exhibits symptoms of electric action; when this has arrived at the maximum of its scale, Henley's is found to be affected, and the record of these three will give the force of electricity of the air under all circumstances. R is a discharger; or, as it is termed in the Greenwich observations, a Ronalds's spark-measurer. The length of the spark is measured by means of a long index, which exhibits the distance of the two balls *x* and *y* from each other on a multiplying scale, *y* being connected with a rod which is raised and lowered by means of the glass lever, *z*. Each division of the scale represents one twentieth of an inch in the length of a spark; the divisions, of course, are not equal, and they serve to estimate fortieths of an inch, or even less.

The results of Mr. Ronalds's observations have been, from time to time, published in the Reports of the British Association for the Advancement of Science. In that for 1849 will be found an elaborate discussion by Mr. Birt of all the observations taken at Kew through a period of several years; to abridge this paper with advantage would be inconsistent with the plan of this work, but reference may be easily made to it by those who wish further information on this very important subject.

A very efficient apparatus, on the model of those instruments in the Kew Observatory, has been constructed at a comparatively small expense; the whole, mounted on a tripod stand, very light and well adapted for transport, may be protected by a covering of wood of sufficient height to allow the observer to stand upright, and of such a breadth as to allow him to walk round the apparatus*.

155. **Peltier's Electrometer.**—The instrument with which M. Quetelet's observations were made, is M. Peltier's induction electrometer; it is por-

* See Brit. Assoc. Report, 1851.

table, being of small size, simple in its construction, certain in its results, and any number may be made perfectly comparable with each other. The electricity obtained from the air is made to deflect a magnetic needle, and the arc of deflection shows the intensity, the value of each degree having been previously ascertained. Though this instrument was exhibited by Prof. Wheatstone at the meeting of the British Association in 1849, and described by him in the 'Report,' I am not aware that it has been used in England; yet from the report of its performance, it would seem well to deserve the attention of meteorologists.

OZONE.

156. Nature of Ozone.—In the year 1848, Dr. Schönbein, of the University of Bâle, discovered a new principle to which he gave the name of ozone; and as observations, with reference to the amount which may be presumed, on the application of the proper test, to exist in the atmosphere, are now taken extensively in England, and still more so in Germany, it demands explanation in this place. Its name is derived from the peculiar smell which distinguishes it, when produced artificially by the electrifying machine.

Of the nature of ozone, as it exists in the atmosphere or as it may be produced by electricity or chemical combinations, various opinions are held. Dr. Faraday considers it to be oxygen in an allotropic state, that is, with a capability of immediate and ready action impressed upon it; its discoverer is disposed to view it as a bin-oxide of hydrogen; as yet, the mode by which oxygen passes into ozone is inexplicable.

157. To procure Ozone.—To procure ozone, let a piece of newly-scraped phosphorus, half an inch long, be put into a two-quart bottle containing just water enough to cover it to the height of a quarter of an inch; raise the temperature of the water to 63°, and ozone will be produced in considerable quantity in the course of five or six hours; the bottle meanwhile being kept closed, but not very tightly, lest the phosphorus should inflame and burst it. The bottle must then be rinsed with water to dissolve the phosphorous acid, and the air that is left will be stongly impregnated with ozone, which may be recognized by a peculiar smell—similar to, but more powerful than, the electric odour. Ozone is produced also when electricity is discharged into the air from a powerful machine by a moist wooden point; there will be a feeling as of a current of vapour escaping, and the test, to be presently described, will show the presence of ozone, and at the same time the peculiar odour is recognized. Now, as electricity may be shown to be generally more or less present and active in the air, it is presumed to be the agent by which oxygen is converted into ozone, or, to seize Dr. Faraday's idea, by which oxygen is rendered more energetic in its action. This, we shall see, has a most beneficial effect in purifying the atmosphere; for ozone unites most readily with fœtid gases and miasmata, which are thereby deprived of their deleterious qualities. Hence the importance to the meteorologist and the physician of studying its properties; it may be that we may light upon some combinations which may render us competent to reduce at will to innocent compounds the most deadly effluvia.

158. Test for Ozone.—The test for ozone may be thus prepared: 200 parts of water, 10 of starch, and 1 of iodide of potassium, are to be boiled together for a few seconds; bibulous paper is dipped into the solution, and then dried. If a strip of this paper is exposed to an atmosphere suspected to contain ozone, it will, should ozone be present, assume a brown tint, which will be turned to blue when the paper is dipped in water; the amount may be judged of by the time it is exposed compared with the depth of the tint. The ozone seizes on the potassium, and leaves iodide of farina, indicated by the blue colour.

159. **Schönbein's Ozonometer.**—Dr. Schönbein's *ozonometer** consists of twelve bundles of paper, prepared with iodide of potassium and starch; each bundle contains sixty strips, and serves for one month's observations; a spare set is added for additional observations during thunder-storms, or whenever the air may appear to be overcharged with electricity. At nine o'clock every morning, a strip of the prepared paper is to be suspended in a spot to which the air has free access, but not the sun. It must be removed from dung-heaps, stables, &c., where gases are developed, which would vitiate the observation. At nine o'clock in the evening the exposed strip is dipped in water; it will be found to assume a purple tint. The depth of this tint is compared with the corresponding colour on a scale, on which there are ten gradations, and the number is to be inserted in the register with which it agrees in depth. Another slip of paper must be exposed at 9 P.M., and examined and registered in like manner at 9 A.M. on the following morning. At the close of each month, the mean is to be deduced, by dividing the sum of the numbers registered, by the number of observations.

160. **Properties.**—The properties of ozone are—

* To be obtained of the agents, Casella and Co., Meteorological Instrument Makers, 23 Hatton Garden.

the peculiar odour, resembling, when the ozone is diluted, the electric smell—when concentrated, that of chlorine; animals expire when placed in it; and it has the effect of rendering respiration difficult and producing catarrhal effects on the human subject. It is insoluble in water; it discharges vegetable colours like chlorine; acts most powerfully on metallic bodies, producing a very high degree of oxidation. It decomposes rapidly phosphuretted and sulphuretted hydrogen gas; this property is the one which combined observations will, it is hoped, tend to develope. We know that immense amounts of deleterious gases arise from the decomposition of animal and vegetable matter. Ozone, even at a low temperature, combines energetically with these, and neutralizes their effects. Schönbein has proved by experiments, that air containing $\frac{1}{8000}$ of ozone can disinfect 540 times its volume of air produced from highly putrid meat; that is to say, such a fœtid atmosphere may be completely purified by a quantity of ozone equal to $\frac{1}{3240000}$ of its volume. Now, in bad localities, it is evident that we may expect the test to show little or no ozone, while, as Faraday found at Brighton, the pure air from the ocean abounds with it. Schönbein met with it in abundance during a storm on the Jura, and could even recognize its smell; so that the purification of the air by storms would seem now to be philosophically proved. The electric discharge, of which thunder and lightning are the sensible indications, produces in large quantities this valuable disinfecting agent.

161. Ozone Observations.—At an interview with Dr. Schönbein, at Bâle, in 1853, I engaged to bring the subject of ozone observations before the meteorologists in this country, for which purpose I addressed a circular to several meteorologists; and the readiness with which the matter was taken up, especially by medical men, showed a deep interest in it. Many observers undertook, for twelve months, simultaneous observations daily, at times which coincided with those taken throughout Germany; the whole were forwarded to me for transmission to Dr. Schönbein.

I cannot say that the examination of these reports forwarded to me, have led me to deductions of any value to science; my own experience, confirmed by that of other observers, shows a want of accordance between observations taken at places even at a few hundred yards from each other. In some large towns, as London and Manchester, no ozone is developed; and it has sometimes happened that the maximum of discoloration of Schönbein's test paper takes place some time before that fixed for registering, and the colour has partially faded before the amount has been recorded.

162. **Moffatt's Ozonometer.**—Dr. Moffatt, of Hawarden, Cheshire, has paid great attention to the subject for many years; his test paper is suspended within a large box from which light is excluded, but which is perforated at the bottom for the admission of air; any arrangement, however, will do that shelters the paper from wet and from the sun's rays; a convenient one is Sir J. Clarke's 'Ozone cage,' made of wire-gauze; the amount of ozone is determined by the strength of discoloration in a given time without immersion in water. Schönbein's tests after immersion in water are valueless; Moffatt's papers, not being moistened, if kept in the dark or between the leaves of a book, may be retained for years, and it has lately been shown that they are more sensitive than Schönbein's; they are consequently recommended by the Council of the British Meteorological Society. In the Report of the Council of the British Meteorological Society for 1854, Dr. Moffatt gave various deductions on the subject of ozone from his own observations, which are deserving of consideration; indeed the whole investigation is worthy of successive observations and experiments, especially in positions far removed from the gases which infect the stratum of air near the ground.

It is contemplated to erect, near the Kew Observatory, a lofty mast, on the summit of which the effect of the air on the ozone test paper may be supposed to be beyond the reach of local influences.

PART III.

PRESENT STATE OF METEOROLOGICAL SCIENCE IN ENGLAND.

Several causes have of late years combined to direct attention, in this country, to the difference of climate, the amount of rain, and other atmospheric phenomena, on which information can be supplied only by the unassuming labours of practical meteorologists. The presence of cholera in England during the year 1849 led many scientific men, especially the members of the medical profession, to inquire whether, during that unhealthy season, any deleterious changes were traceable in the conditions of the atmosphere. The removal of protective laws from the produce of native agriculture has compelled the farmer to call in the aid of science to increase the quantity and improve the quality of his crops, the confined limits of our small island opposing his extending his operations; hence he has anxiously sought information as to the mean temperature of his locality, in order that he might commit to the soil those productions only to which the climate may be considered favourable. It has followed that the science of meteorology, before confined to students of natural philosophy, has been favourably viewed by the agricultural class of the community, and the pages of their journals are open to communications on that subject.

163. British Meteorological Society.—At the commencement of the year 1850, a small number of practical meteorologists, whose attention to the science was not of sudden growth, were induced to take into consideration the possibility of enlarging their sphere of operation—of collecting facts and observations in such number as to form the groundwork for generalization.

The result of their deliberations was the formation of the British Meteorological Society, which now numbers some hundred members, a large portion of whom are practical meteorologists. They have sent out some valuable reports, and have decided upon a form of registration which includes columns for all phenomena likely to come under observation. Those who wish to unite with the established corps of observers would do well to communicate with the Secretary, James Glaisher, Esq.*, whose services in the cause of meteorological science are well known. One of the most important steps taken by the Society has been the verification of instruments

* 13, Dartmouth Terrace, Lewisham, Kent.

and the comparison of their readings under similar circumstances; the object being to remove all sources of error arising from imperfection of barometers and thermometers in use among the members.

164. Registrar-General's Reports.—For some years past the Registrar-General has subjoined to the published weekly returns of births and deaths in London, the results, for the week, of meteorological observations taken at Greenwich. These include the mean daily readings of the barometer, thermometer, and hygrometer; the difference in temperature of each day from the mean of the preceding forty-three years; the force and direction of the wind; the amount of rain; notices of the electric state of the atmosphere; remarks on the amount of cloud, on sudden changes in temperature, on the strength of the wind, or any other phenomena deserving of notice; the indications of a thermometer exposed to the full rays of the sun; the reading of another sunk 2 feet below the surface of the river Thames. As these papers have an immediate and extensive circulation, and as they classify the causes of death by referring each to its peculiar class of disease, they supply the means of comparing the prevalence of any one in particular with any peculiarity in the state of the air. At the end of the year a digest of the whole is regularly published on a single sheet, for facility of reference.

The weekly returns from the Observatory at Greenwich are incorporated, every three months, with a digest of all the returns of births, marriages, and deaths which have been received during that time. The meteorological portion of this Quarterly Report embraces, beside, returns from about fifty places scattered all over the country; and such has been Mr. Glaisher's care in the comparison of instruments and systematic training of all the observers, that considerable confidence may be placed in the results. As this is the most valuable combination of observers in the science of which this country can boast, it may be proper to explain the grounds of that confidence which they may fairly claim.

The individuals who have undertaken the observations are, with few exceptions, either graduates of one of the Universities, or fellows of some learned Society; it may therefore be presumed not only that they are competent to record phenomena, but that their character would impress the returns with a stamp of trustworthiness and authority. Their barometers have been constructed by the best makers, and have been compared with the Royal Society's flint-glass standard, either directly or intermediately. The scales are of brass, leading from the cistern throughout the whole length of the tube. The indices read to ·002 of an inch by means of the vernier, and by estimation to ·001. Before the results are forwarded to Mr. Glaisher, the barometric readings are all reduced to one standard temperature, viz. 32° of Fahrenheit, the freezing-point of water. The wet-bulb and dry-bulb thermometers have also been compared with standards, and from simultaneous observations with them, are deduced the dew-point, the tension of aqueous vapour, the degree of humidity, the weight of vapour in a cubic foot of air, the weight of vapour requisite to complete the saturation of a cubic foot of air, and the weight of a cubic foot of air of the mean temperature and density of the month.

The thermometers indicating the greatest and least temperature occurring in the preceding twenty-four hours, are registered at 9 A.M. They are for the most part of Rutherford's construction; but for the maximum reading, Negretti and Zambra's maximum thermometer is now preferred.

The rain is measured at 9 A.M., and the quantity recorded is the produce of the preceding twenty-four hours. The force of the wind is estimated by some observers from the indications of Dr. Lind's anemometer. All concur in recording, as nearly as possible, the approximate value to be given to the amount, by reckoning a gale as 6, a calm as 0. The nomenclature of the clouds is that first proposed by Luke Howard, Esq. A clouded sky is represented by 10, and a clear sky by 0; the interpolations are arrived at by estimation. At most of the places, too, a register is kept of the amount of ozone.

Though some parts of England are still without a representative, the positions of observers are tolerably well distributed. Thus then we may conclude, I apprehend, fairly, that well-founded reliance may be placed on the monthly reports of those gentlemen whose names are appended to them, on every ground, whether we regard their position in society, the valuable instruments in which they have invested no small outlay, or the unison of action which characterizes their proceedings.

The three months' observations are forwarded to Mr. Glaisher, with every particular requisite to reduce them to what they might be supposed to have been had they been taken at the level of the sea. They are then arranged in groups by him, according to the latitude, and from them are deduced certain results regarding the climate of the various parts of England, coincidences or irregularities of atmospheric phenomena, and of natural occurrences, such as the arrival and departure of migratory birds, the time of flowering of plants, the progress and prospects of agriculture, falls of snow, thunder-storms, appearance of meteors and auroræ, on all which subjects the observers are expected to report for their own localities.

Mr. Glaisher's Quarterly Reports, diffused throughout the community by means of the newspapers and the scientific periodicals, have been the means not only of spreading information valuable to the man of science, the physician, the agriculturist, and the engineer, but of exciting and keeping alive an interest in the study of atmospheric phenomena. The labour of comparison and reduction is very great, and that gentleman has received, as he has well deserved, the thanks of all those who desire the advance of science, for his indefatigable labours in the cause.

165. Conference at Brussels.—In consequence of certain representations made to the British Government by that of the United States, a conference was held at Brussels in August 1853, to bring before representatives from the maritime states a plan which had been submitted by Lieut. Maury, of the United States Navy, for extending the field of research into the laws which govern the circulation of the atmosphere and control the currents of the ocean, by combining the marine of all nations in one uniform system of observation. The methods of observation, the character of the instruments, the scales to be adopted, all underwent ample discussion; the points recommended by the Conference for the adoption of their respective governments were:—

α. An improvement in the construction of instruments, especially the Barometer and Thermometer, over those inferior instruments hitherto found on ship-board.

β. A comparison, *inter se*, between the instruments used by all the countries whose representatives were present at the Conference.

γ. An 'abstract log' was agreed upon, to be filled up systematically by all the captains in the navy of each nation, and as many captains in the merchant service as should volunteer their co-operation.

δ. A depôt was recommended to be established in each country for the joint contributions of every ship furnished with the suitable instruments, so that the whole set of observations may, from time to time, be discussed, and the results rendered available for future navigators*.

The 'abstract log' requires observations at least

* Report presented to Parliament and ordered to be printed, 8th of February, 1854.

five times daily—but recommends fourteen—of the ship's place; direction and rate of currents; observed magnetic variation; the direction and force of the wind, on Admiral Beaufort's system of registration, which is best adapted for nautical persons (though in the papers published by the Board of Trade the Greenwich notation, 0—6, will henceforward be used); the height of the barometer; the dry- and wet-bulb thermometers; the forms and direction of clouds; the proportion of sky clear; the hours of fog, rain, snow or hail; the state of the sea; the temperature at the surface; the specific gravity and temperature at any depth of the water of the ocean; the state of the weather;—and an ample column is left for remarks, the subjects suggested for which are tempests, tornadoes, whirlwinds, typhoons or hurricanes, waterspouts, temperature of rain, shooting stars, auroræ, halos, rainbows, meteors, appearance of birds, insects, fish, sea-weed, or drift-wood out at sea, and tidal observations;—thunder, lightning, and electrical phenomena.

The high scientific standing of those who assisted at the Conference, is a sufficient guarantee for the importance to science of these objects of research, which have been thus enumerated in full, as suggestive to observers in general of the kind of phenomena worthy of registration.

The first movement of this truly noble assembly was to elect M. Quetelet of Brussels, the representative of Belgium, their President; and at thirteen consecutive meetings, at which from the report it is evident that much important business was got through, the entire subject was discussed in detail, and decisions were arrived at only after the most mature consideration.

166. Recommendation adopted by Government.—The British Government, following up the recommendation of this Conference, has created a special department of the Board of Trade to carry out the objects proposed, under the superintendence of Rear-Admiral FitzRoy, who thus explains the object in view:—"All the valuable meteorological data which have been collected at the Admiralty, and all that can be obtained elsewhere, will be tabulated and discussed in this new department of the Board of Trade, in addition to the continually accruing and more exact data to be furnished in future. A very large number of ships, chiefly American, are now engaged in observations, stimulated by the advice, and aided by the documents so liberally furnished by the United States Government, at the instance of Lieut. Maury, whose labours have been incessant. Not only does that Government offer directions and charts gratis to American ships, but also to those of our nation, in accordance with certain easy and just conditions. In this country, the Government, through the Board of Trade, will supply a certain number of ships which are going on distant voyages with 'abstract logs' (or meteorological registers) and instruments gratis, in order to assist effectively in carrying out this important national undertaking. In the preface to a late edition of Johnston's 'Wind and Current Charts,' published last June at Edinburgh, Dr. Buist says,—'It has been shown that Lieut. Maury's charts and sailing directions have shortened the voyages of American ships by about a *third.* If the voyages of those to and from India were shortened no more than a *tenth,* it would secure a saving, in freightage alone, of £250,000 annually. Estimating the freights of vessels trading from Europe with distant ports at £20,000,000 a year, a saving of a tenth would be about £2,000,000; and every day that is lost in bringing the arrangements for the accomplishment of this into operation, occasions a sacrifice to the shipping interest of about £6000, without taking any account of the war navies of the world.' It is obvious that, by making a passage in less time, there is not only a saving of expense to the merchant, the shipowner, and the insurer, but a great diminution of the risk from fatal maladies, as instead of losing time, if not lives, in unhealthy localities, heavy rains, or calms with oppressive heat, a ship properly navigated may be speeding on her way under favourable circumstances. There is no reason of any insuperable nature why every part of the sea should not be known as well as the land, if not indeed better than the land, generally speaking, because more accessible and less varied in character."

167. Instruments supplied.—One of the first steps taken by the Board of Trade in furtherance of the object recommended by the Brussels Conference, was to apply for advice and assistance to the Kew Committee of the British Association for the Promotion of Science. Under their auspices a cheap thermometer and barometer have been constructed, which not only our Government, but that of the United States, have adopted.

Messrs. Negretti and Zambra, and Messrs. Casella and Co., have at the low price (wholesale) of 5*s.* 6*d.*, engaged to supply any number of thermometers on the model proposed by the Kew Committee; they are $10\frac{1}{2}$ inches in length, with a range of graduation from 10° to 130° Fahr.; the tube is enamelled, and the divisions etched on it with fluoric acid; the figures are stamped on a brass scale, or else marked on a porcelain scale by

Messrs. Negretti and Zambra's new process, and the whole is enclosed in a copper case.

One of the conditions of supplying the Government demand was that each instrument should be examined and tested at the Kew Observatory; the numbers in the course of verification there at the latter part of 1854, were, for the United States' Navy 1000 thermometers and 50 barometers, for the Board of Trade 500 thermometers and 60 barometers. I have in my possession one of these thermometers by Casella, numbered 1020 and marked K. O., to indicate its having passed the ordeal, and the very small corrections which it requires (viz. at 32°, $-0°\cdot3$; at 42°, $-0°\cdot1$; at 52°, $+0°\cdot1$; at 72°, 0; at 92°, $-0°\cdot1$) will clearly indicate the important advance toward correct instrumental appliances to which this movement has given rise.

The barometer selected has already been described, p. 209; its price, including the cost of packing case, the brass arm for suspension, and 10*s.* for verification at the Observatory, will be to Government when supplied wholesale, £3 15*s.* 6*d.*; this instrument has also to undergo severe scrutiny, and to be compared with a new standard which has lately been erected in the Kew Observatory. The diameter of the tube of this standard is one inch; the zero-point is brought down to touch the surface of the mercury in the cistern, and the height of the column is determined by two telescopes moving on a vertical brass scale at the distance of some feet from the tube containing the mercury. The point which is brought to the surface of the metal in the cistern is at a known and well-determined distance from a cross above it; in the lower telescope a horizontal wire is made to bisect this cross, and in the upper another wire measures the distance from the cross to the top of the column.

A large receiver, in which the atmospheric pressure is varied at pleasure, is used for comparing the various barometers, the whole being under the superintendence of Mr. Stewart; through the Kew Observatory facilities are granted to private observers, as well as public institutions, for having the errors of their instruments determined, before entering upon a course of registration of atmospheric phenomena. Here also may be obtained, at a moderate cost, standard thermometers which have been divided on the stem with great accuracy, and with the minutest regard to every consideration which may tend to bring the instrument as near as possible to perfection.

The Royal Observatory, Greenwich.

168. For a period of not less than 160 years,

systematic observations of the places of the sun, moon, planets, and fixed stars, have been recorded at the Royal Observatory with an accuracy not surpassed, if indeed equalled, elsewhere. The Government of this country have evinced a sound discretion in the appointment of the most eminent mathematicians to the office of "Astronomer Royal," an office now held by G. B. Airy, Esq., whose mind is ever actively engaged, not only in sustaining the character of the Observatory for accuracy of observation by the introduction of instruments superior to any hitherto employed (witness the new 14-feet transit-circle, and the alt-azimuth circle), but in adding to its efficiency in every collateral branch of science. Under his auspices the magnetical and meteorological observatory was originally established; and of late, the new system of automatic registration of the magnetical and meteorological instruments by means of photography (invented and brought to perfection by Charles Brooke, Esq., M.B., F.R.S.), has been introduced, and now forms a most striking feature in that department of observation. Successive Governments have shown a liberality in promoting the objects for which the Observatory was originally founded, namely to assist the navigator in traversing the pathless ocean.

In the year 1837 it was determined to erect a

magnetic observatory, for the purpose of investigating the laws of magnetism, on the full understanding of which the mariner's compass depends for improvement, and the chart by which the navigator is guided for its accuracy; conjointly with these investigations an elaborate system of meteorological observation was commenced, in the expectation of discovering some of those causes which produce the variations in the conditions of the atmosphere, a kind of knowledge auxiliary to navigation, in which so much depends upon that variable element, the wind. Greenwich, moreover, was understood to be well appointed in a trained corps of observers, renowned for the accuracy and care which they had employed in the most exact science; and the published Reports, which originated in 1840 and have extended to 1851, have proved the wisdom of the choice of that locality for magnetical and meteorological instruments.

The magnetic observatory is a small detached building, its nearest angle being 230 feet from the nearest part of the astronomical observatory, and 170 feet from the nearest outbuilding; the material is wood, and iron has been carefully excluded from its construction; the form is that of a cross, with four equal arms nearly in the direction of the cardinal magnetic points; its extreme length and breadth are each 40 feet, and the breadth of each arm, which is 10 feet high, is 12 feet. The only iron to be found throughout the whole building is in the fire-grate in the ante-room, the mean-time clock, the sidereal-clock, and the check-clock.

Though the magnetical instruments do not form a part of the meteorological establishment of the Observatory, strictly speaking, yet, as their movements are registered photographically, and in combination with those devoted to atmospheric changes, I consider it necessary to say somewhat respecting them, in connexion with those which record more particularly atmospheric variations.

On Photographic Registration of Meteorological Phenomena.

169. On this subject I shall endeavour to be sufficiently explicit to convey a clear notion of the ingenious contrivances by which automatic registration is attained; but it is not my intention to describe them so minutely as would be requisite were I about to give directions for their construction. Those who may wish to adopt a similar apparatus, may consult advantageously the Philosophical Transactions, Part I., 1847, and Part I., 1850; and the Greenwich Magnetical and Meteorological Observations, 1847.

On the first introduction of photographic registration by Mr. Brooke, that gentleman adopted the light of a camphine lamp, as producing the most powerful photographic effect. For this has now been substituted a mixture of common coal-gas and naphtha, which is found to be quite equal in brilliancy, and far more manageable. The gas, on admission into the magnetic observatory, is received into a tin box divided horizontally into two compartments, the lower of which contains water, and half of the other is filled with naphtha. This upper half is partitioned off into eighteen cells by vertical divisions, each attached alternately to different sides of the box.

Fig. 3, Plate VIII. is a section, and fig. 4 a plan of this box or receiver; *c* is the portion partially filled with water, which is heated by the jet of gas *f*; *d* is the naphtha compartment, half-filled with that substance; as the water heats the naphtha, the upper part of the compartment at *e* becomes filled with vapour, and the gas entering at *a* traverses the compartment in the direction of the arrows, as shown in fig. 4, unites with this vapour, and the two gases mutually diffused issue at *b*, and thus combined are distributed throughout the building.

The paper on which the photographic trace is received is a strong woven paper, of equal texture throughout; in manufacturing it, all foreign substances which might combine injuriously with the chemical substances used in its future preparations have been carefully excluded.

A sufficient quantity of paper for the consumption of three or four weeks is treated in the following manner:—"To a filtered solution of 4 grains of isinglass in 1 fluid ounce of boiling distilled water are added 12 grains of bromide of potassium and 8 grains of iodide of potassium. The solution, either when hot or cold, is evenly laid upon the paper with a camel's-hair brush, in such quantity as thoroughly to wet its surface, but not to run off; the paper is then dried quickly before the fire. The paper thus treated is preserved by keeping it in a dry place and in a drawer.

"When a cylinder is to be charged with photographic paper, the room is darkened and illuminated only by a candle, whose flame is surrounded by a cylinder of yellow glass. The paper is laid flat in an earthenware dish, and is washed with an aqueous solution of nitrate of silver, made by dissolving 50 grains of crystallized nitrate of silver in 1 fluid ounce of distilled water, which is laid on in quantity not sufficient to run. The paper is then in a state fit to be placed upon the cylinder.

"When the paper is to be taken off the cylinder, the room is illuminated in the same way, the cylinder is detached from its mounting, the external cylinder is drawn off, and the paper is unfolded and laid flat in a dish. In this state it exhibits no trace of the action of the light. It is then washed with a solution of gallic acid, to which a few drops of acetic acid are added, till it is moderately wet all over; the impression begins soon to appear, and in a few minutes acquires its full strength. The paper is then repeatedly washed with water till the water runs off quite clear. Solution of hyposulphite of soda (formed by dissolving 1 drachm of the hyposulphite in 5 ounces of distilled water) is then poured upon it, and water is added in considerable quantity; after this has remained about five minutes, the paper is washed repeatedly with water. The trace is then securely fixed, and light may be admitted into the room. The sheets are then usually preserved for gradual drying within the folds of linen towels."

The cylinders alluded to in the above extract are those around which the paper is wrapped to receive the photographic trace. They are, in fact, French glass-shades (such as are used to protect works of art), 11½ inches in length, and 14½ inches in circumference. The shade, after having been blackened in the inside, is cemented into a cap 1 inch deep, having a brass pin projecting from the centre. A second shade, a little larger than the former, is placed over the paper when it has been attached in a moist state to the first; this latter cylinder is kept in its place by a few turns of tape round the collar part, which is moistened with water; damp list is also placed between the hemispherical parts of the shades. This provision is necessary to prevent the paper from becoming dry during the time it is subject to the photogenic action, for dryness would very materially its sensibility. When the axis of the cylinder is required to be horizontal, as in the registration of horizontal movements, the pin which is in the line of the axis, and the cylinders themselves, rest on friction-rollers; a bent wire on the axis is caught by a fork attached to the hour-hand of a time-piece, which is about the size of a ship's chronometer, and thus the cylinder is carried round once in twelve hours, or any other period which may be determined, with such smoothness and ease as not to alter the rate of the time-piece in the slightest degree. One-twelfth of the circumferenee of the cylinder will evidently measure one hour, and about $\frac{1}{10}$th of an inch will be the measure of five minutes of time.

Fig. 1, Plate VIII. *a*, vertical cylinder charged

with photographic paper; *b*, wooden cap; *c*, central pin.

Fig. 2, *d*, the paper unwound and divided into twelve parts, marking the hours; *e*, *f*, the trace of the movement of the mercury in the barometer during that time; *g g*, (*g′ g′* in fig. 1), photographic base-line.

The time-piece, in moving the vertical cylinder, lies flat underneath it. In the case of a horizontal cylinder it is placed with its face vertical, and facing the cap (see fig. 5, B).

For the sake of convenience, each cylinder is made to perform double duty. The barometer and the vertical-force magnetometer are registered on the same cylinder, and their traces are allowed to cross each other in opposite directions, which, with a careful adjustment, can easily be effected without interference. The declination and horizontal-force magnetometers are registered in the same way, on a cylinder whose axis is horizontal; and the dry- and wet-bulb thermometers share a cylinder between them.

To describe now the manner in which the photographic trace is left, commencing with the declination magnet. The light by which the trace is made, is placed slightly out of the direction of a straight line joining the suspension-skein of the magnet and the centre of the photographic sheet. The chimney which covers the light (a jet of gas united with the vapour of naphtha) is perforated by a slit $\frac{3}{10}$ths of an inch long and $\frac{1}{100}$th of an inch broad (see fig. 5, *d*); the light from this slit falls on a metallic concave mirror (*e*) which is carried by the suspension apparatus of the magnet, and moves with it; by it the light is made to converge about the centre of the cylinder of photographic paper, at a distance of nearly 12 feet. To reduce the image of the slit to a neat spot of light, a cylindrical lens of glass, *b*, is interposed. Now, as the magnet, and with it the mirror, turns in azimuth, the image of the slit runs along this lens; and, at whatever part it falls, it is concentrated into a definite and brilliant spot of white light, which leaves a photographic trace on the prepared paper; and as this is constantly carried round by the time-piece, the effect produced will be a continuous line around the cylinder, *c*, with deviations to the right or left indicating the horizontal movement of the magnet.

As in practice it is found that the length of the paper is not always the same, it is therefore necessary to have a time-scale for each portion after it has been detached from the cylinder; this is effected by shutting-off the light for an instant, which causes a break or light space in the photographic trace. The time is noted accurately, and the same thing is repeated, we will suppose, one or two hours afterwards; the distance between these breaks supplies data for a time-scale for that special register.

To measure the ordinates from this time-scale, which may be considered as a line of abscissæ, the actual deviation of the magnet at particular instants, four times daily, is read off by a theodolite (carefully adjusted for the purpose) in degrees, minutes, and seconds of arc; these readings, compared with the length of the ordinates at those times, supply the means of reducing all the others to the same standard.

As it can never be expected to obtain glass cylinders with perfectly cylindrical surfaces, or perfect surfaces of revolution, there is a probability that the line of intersection of a plane perpendicular to the axis of the cylinder with the paper on the surface, will not be a perfectly straight line when the paper is opened out. To obtain a base-line on each sheet the following plan is adopted:—An independent ray of light, impinging perpendicularly to the axis of the cylinder from a light 6 inches distant, is received by the cylindrical lens, and marks a strong line all round the cylinder, which, when the paper is unrolled, becomes the line of abscissæ on which the times are set off; while perpendicular ordinates from it will be proportional to the movement which is the subject of measure (see *g g*, in fig. 2).

The arrangements for the horizontal-force magnet are precisely the same as those described for the declination magnet. Every part of the cylinder apparatus, except that on which the light falls, is covered with a double case of blackened zinc, having a slit on each side on the same horizontal plane as the axis of the cylinder; and every part of the path of the photographic light is protected by blackened zinc tubes from the admixture of extraneous light.

The vertical-force magnet traces its line of movement by reflected light on a cylinder charged with paper, whose axis is vertical; the other portions of the apparatus resemble so nearly that already described that a further account is unnecessary. On the east side, the same cylinder receives the trace of the barometer. At the distance of 30 inches is a large siphon barometer, the bore of the upper and lower extremities of its arms being about $1\frac{1}{10}$ inch; a glass float in the quicksilver of the lower extremity is partially supported by a counterpoise acting on a light lever (which turns on delicate pivots), so that the quicksilver constantly supports a definite part of the weight of the lever. This lever is lengthened (see fig. 6), to carry a vertical plate of opaque mica with a small aperture, whose distance from the fulcrum is so regulated with regard to the distance of the point of action of the float-wire, that its movement is four times the movement of the column of the cistern-barometer. Through this hole the light of a gas-jet, collected by a cylindrical lens, shines upon the photographic paper. Another pencil of light from the same jet shines through a fixed aperture, with a small cylindrical lens, for tracing a photographic base-line upon the cylinder of paper, similar to that for the cylinder of the declination magnet.

Such parts of the apparatus adapted to photographic registration of the declination magnet and barometer only are shown in the engraving, as are requisite to explain the mode of its action.

In fig. 5, A is the cylinder covered with photographic paper, the axis of which is horizontal; B is the time-piece which gives it a rotatory motion; *b* is a cylindrical lens, bringing to a point the light from the jet *d*, which has been reflected by the mirror *e*; *f* is the magnet suspended by the silk thread *g*; as this turns in azimuth, the mirror *e* turns with it, and the reflected image of the slit in the chimney covering the jet of light runs along the cylindrical lens, by which it is brought to a point on the paper, on which it leaves a trace, as shown at *c*. In the course of one rotation of the cylinder this trace will have gone all round it, with deviations to the right or left indicating the movement of the magnet in azimuth during the time occupied by the rotation.

Fig. 6 shows the arrangement of the barometric apparatus. Q *e* is a lever whose fulcrum is *e*, the counterpoise *f* nearly supporting it; *s* is an opaque plate of mica, with a small aperture at *p*, through which the light passes, having before been refracted by a cylindrical lens into a long ray, the portion only of which opposite the aperture *p*, impinges on the paper; *d* is a wire supported by a float on the surface of the mercury; G, H is the barometer; P, the vertical cylinder charged with photographic paper; *r*, the photographic trace; I, the time-piece, carrying round the cylinder by the projecting arm *t*.

It is evident that the respective distances of the float and the aperture *p* from the fulcrum may be regulated so that the rise and fall of the float may be multiplied to any extent required. At Greenwich, the extent of the photographic record is four times the actual rise and fall of the mercury in the cistern. These contrivances were shown in the Great Exhibition of 1851, Mr. Brooke having supplied his apparatus.

The dry- and wet-bulb thermometers are registered by the same means as the instruments already described. They are very large, for thermometers of the usual size would not sufficiently shut off the light. The fluid employed is quicksilver, and the bore of the tube is $\frac{4}{10}$ths of an inch; the tube is cylindrical, and 8 inches long; the bulb of the wet-bulb thermometer is covered, in the usual way, with muslin, to which moisture is communicated by the capillary passage of water through lampwicks. They are capable of elevation by means of a coarse screw, so that the mean temperature for the time of observation may be brought near the centre of the cylinder; but the bulbs are so adjusted as to stand about 4 feet from the ground, the small variation in height being simply for the purpose of having the trace recorded upon a convenient part of the paper. Plates cover the thermometer-frames, with apertures so narrow that the column of mercury shuts out the light. Across these plates a fine wire is placed at every degree, and a coarser wire at every 10°, and also at 32°, 52°, and 72°, so that there may be no chance of mistaking the reading of the degrees of temperature. The light of a jet of gas is condensed by a cylindrical lens whose axis is vertical, into a well-defined line of light, which shines through the thermometer-stalk upon the cylinder of paper, which is close to it. As the cylinder of paper revolves under this light, it leaves a broad sheet of photographic trace, the breadth of which varies with the varying height of the quicksilver in the thermometer-tube; but, inasmuch as the light is intercepted by the wires placed across the tube at every degree, there are spaces traced by the wires in which there is no photographic action. These appear on the paper in the form of light lines on a dark ground, and serve the purpose of reading off the thermometers, which is facilitated by the broader lines marking the decades of degrees; nor is any photographic base-line needed, for the wires form the only register required. The cylinder receives the trace of the wet-bulb on one side, and of the dry on the other. Its axis is of course vertical, and it is made to revolve once in forty-eight hours; the paper, when removed, will therefore show the variation of both thermometers during the last twenty-four hours, one-half of the photographic trace being due to the dry-bulb, the other to the wet-bulb thermometer. The circumference of this cylinder is 19 inches.

Such, then, are the arrangements for the automatic registration of meteorological and magnetical instruments now introduced into the Greenwich Observatory, and their value, as indicating the minutest movements, is very great; while the labour of watching each instrument and recording its variations every two hours is entirely dispensed with. The consequence of the introduction of self-registration has been that two observers are more efficient than four under the old system. We are not aware, as yet, of the effect of time on the photographic trace; to ensure permanency, therefore, the variations of the instruments are inked in by a definite line along the edge. The papers are kept carefully arranged in the daily order, and ready for immediate reference, with the other records of the establishment.

Of the radiation thermometers, which measure the amount of heat radiated from the earth's surface, of those sunk beneath in the soil, of the thermometer 2 feet below the surface of the river Thames, and of the actinometer, which measures the direct heat of the solar rays, little need be said; I therefore pass on to the

Anemometers.

170. To have the means of registering the amount and direction of the wind for every hour of the day had long been a desideratum with scientific men, and much ingenuity has been shown in the mechanical contrivances which have been entered upon for that purpose. The instrument which has met with the greatest approba-

ıgland is Osler's anemometer, one of been erected lately at the new Royal Exchange, and another has been in use at Greenwich for many years; its indications are constantly recorded and are considered by competent judges to be very trustworthy, the instrument having undergone various changes and improvements since its first erection. The instrument traces on a sheet of paper the direction and pressure of the wind, and the amount of rain which may have fallen in twenty-four hours. A copy of this register is shown in Plate IX. The anemometer itself consists of a vane V, turned by the wind, attached to a hollow vertical spindle WX: the paper is divided longitudinally by lines, the central showing the direction of the wind marked S., W., N., E., S.; the upper part receives the trace which indicates the amount of rain; the lower part shows the amount of pressure of the wind on a square foot of surface exposed to its full force. The register paper is placed on a board M, and accurately fixed every day at 10 A.M. This board is carried along by the clock shown at C, at the rate of about an inch per hour. The engraving shows the original contrivance to effect this object, but in consequence of the continual failure of this chain-apparatus, another construction has been adopted; the movement of the board has now been effected at Greenwich by rack-work [illegible] the pinion of a clock.

The pencil 1 is the index of direction [illegible] pencil is operated upon by the vane V, turning [illegible] hollow spindle; there is a pinion at *r*, which, as the vane turns in the direction of the wind, acts on the rack-work of a transverse bar, *e f*, and so causes it to move on the one side or the other.

The centre of the board lies due north and south; if, therefore, the wind blows from the north for twenty-four hours, it is evident that the trace will be along the centre of the board throughout its whole length; if the wind at a certain time veers to the east, the transverse board, and with it the tracing-pencil 1, will be turned aside by the action of the pinion in the cogs, and the line now described will be parallel to the direction of the other, at a distance from it equal to one-fourth of the number of cogs which would come into action at an entire revolution of the spindle; the trace in this direction will continue till the wind again shifts, and the number of horary divisions through which it extends will show the time during which the wind was blowing from that quarter.

The first adjustment for azimuth was obtained by observing, from a certain point, the passage of a star behind the vane-shaft, and from that observation computing the azimuth; then, on a calm s drawn by a cord to that position, …k was so adjusted that the pencil's … on the sheet corresponded to that azimuth.

…or the pressure of the wind the shaft of the vane carries a plate 1 foot square (T, in Plate IX.), which is supported by horizontal rods *n*, *m*, sliding in grooves; this plate is urged in opposition to the wind by three springs enclosed in the box *t*, so arranged that only one comes into play when the wind is light; and the others necessarily act in conjunction with the first as the plate is urged more and more forcibly by the wind. A cord from this plate passes over a pulley and communicates with a copper wire running down the centre of the spindle, which is finally brought to pull upon the spring-lever *v*, and thus the pencil 2, which is attached to it, is drawn in a direction transverse to the motion of the board, the further from its zero line in proportion to the force with which the plate is driven back by the wind. A series of lines numbered 2, 4, 6, 8, &c., shows the amount of pressure on the square foot; the intervals of these lines are adjusted by applying weights of 2 lbs., 4 lbs., 6 lbs., &c., to move the pressure-plate in the same manner as if the wind pressed it. The pin 3 registers the amount of rain, which is thus recorded. The water which has been collected by the gauge passes into the vessel *a*, which is supported by spiral springs *b, b,* which shorten as the quantity increases; into the bottom of this vessel is fixed a tube *c,* open at both ends, in a vertical position, over the top of which is loosely placed a larger tube *e,* closed at the top; when the water has risen to the level of the inner tube it begins to discharge itself gradually into a tumbling bucket *d,* which is enclosed in a globe under the receiver; when full, the bucket falls over and discharges its contents, which run through the waste-pipe *f,* and cause an imperfect vacuum in the globe, sufficient to produce a draught through the pipe *c,* which thus acts as the longer leg of a siphon, and the water continues to flow from the receiver through the interval of the two pipes *c* and *e* till the whole is drawn off, when the spiral springs *b, b* immediately elevate the receiver to its original position.

Now, if we suppose the quantity of water necessary to produce the action thus described to be equivalent to one-fourth of an inch of rain, the mode of registration will be easily understood. The pin 3 is connected by means of the cord *g, g,* with the receiver, which cord is kept tightened by the spring *h*; as the apparatus descends from the weight of water during the fall of rain, this pin advances further and further from the zero of the scale which is shown upon the registering paper, until a quarter of an inch has fallen, when, as this is drawn out and the receiver ascends, the pin is drawn back to its original position, and the same process is repeated.

The register represented in Plate IX. is supposed to record the phenomena of twenty-four hours. It will be seen that rain continued to fall for nearly four hours, when, a quarter of an inch having been received, the trace was brought back to the zero line; five hours afterwards another quarter of an inch had been collected; in two hours more about two-tenths of an inch, when the rain ceased, and none fell for four hours, as is denoted by the line traced parallel to the zero line; rain then fell for an hour; a cessation of three hours followed; two hours after another quarter of an inch was collected; and, for the remainder of the time, a gentle fall is indicated by the gradual departure of the trace from the zero line; the amount of rain collected in the twenty-four hours will therefore in this case be $\cdot25+\cdot25+\cdot25+\cdot06=\cdot81$ inch. On the same paper the traces of the force and direction of the wind may be seen and readily understood. The point of the compass from which the wind blew at any hour is registered along or near the centre of the paper, and the force at the lower part; the zero being the bottom line, and the increase of force being indicated by the departure of the trace from this line towards the inner portion of the paper.

These explanations serve to exhibit the general principles on which this beautiful apparatus is constructed, though the details may occasionally differ. The anemometer and pluviometer have been many years in use at Greenwich, and their registrations are considered very satisfactory. The noble building of the Royal Exchange has been supplied with an anemometer on the same construction, except that the register is vertical; and the anxious merchant, by inspecting the register, can easily satisfy himself whether the wind of the preceding night or day has been favourable to the arrival of some richly-laden vessel, of which he may be in daily expectation.

171. Whewell's Anemometer.—Another anemometer, invented by the Rev. Dr. Whewell, Master of Trinity College, Cambridge, is likewise in constant action at Greenwich; it is also self-registering, and indicates the rate of movement of the air and the directions in which that movement takes place. A horizontal brass plate (Plate X.) is connected with a vertical spindle, which passes through the axis of a fixed cylinder, having a vertical bearing upon a plate at the bottom of it, and a collar bearing in a horizontal plate at the top of the cylinder. The vane, V, turns the whole of the apparatus above the cylinder, which consists of a fly, F, and a system of wheels working into each other; as the fly turns round, with greater or less rapidity according to the motion of the air, these wheels are set in action, and communicate motion to a vertical screw 15 inches in length; the revolution of this screw causes a pencil, P, which is connected with a nut, to descend. The cylinder, C, is covered with paper, on which are marked the points of the compass, and on this the pencil leaves a trace, the length of which is proportioned to the force of the wind. The fly has eight sails, like those of a windmill, inclined at an angle of 45° to the direction of the wind; upon the axis is an endless screw, which works a vertical wheel, of 100 teeth; another endless screw on its axis works a horizontal wheel, of 100 teeth, which is attached to the great vertical screw, S, by which motion is given to the pencil; the descent of this pencil is measured by a vertical scale, and a calculation is made, from accurate measurements of the different parts of the apparatus, of the amount of horizontal movement of the air which is due to an inch of the screw's downward movement. The following are the measures of the principal parts of this anemometer:—

Length of each sail from axis to end ...... 2·30 in.
Length of the flat part of each sail .... 1·92 in.

Inclination of each sail to the wind.........	45°
Forty-five revolutions of the vertical screw correspond to..............................	2 in.
Number of teeth in the vertical wheel ...	100
Number of teeth in the horizontal wheel...	100

Therefore, 10,000 revolutions of the fly cause the pencil to descend through the distance of one thread of the vertical screw, or through a space equal to $\frac{2}{45}$ths of an inch = 0·044 in.

Assuming that the effective radius of the sail is 1·7 in.,—

	in.
The circumference described is 1·7 in. × 2π =	10·68
Therefore the motion of the wind in one revolution is	10·68
In 10,000 revolutions	106,800

corresponding to ·044 in. of the vertical screw, or to one revolution of the screw. From this it follows, that the motion of the wind corresponding to the descent of the pencil through 1 inch is 200,250 feet, or 37·9 miles.

The results of Osler's anemometer give the force and direction of the wind, and those of Whewell's give the amount of horizontal movement in the air, for twenty-four hours; these are amongst the weekly published results of the Greenwich observations. There is also Robinson's anemometer at work at Greenwich.

We have thus then taken a view of the instrumental means and the organization with which, in England, we are provided for the purpose of recording atmospheric phenomena; nor are these all, for at Oxford, under the late Manuel J. Johnson, Esq., of the Radcliffe Observatory, an efficient system of photographic registration was introduced, the arrangements for which differ in several respects from those adopted at Greenwich. The bulbs of the dry and wet thermometers are in the open air, but the tubes are led horizontally through the wall of the building, within which they are bent suddenly in a vertical direction; the mercury, as it rises or falls, cuts off the light which shines through a slit, and leaves a negative trace of the variation in temperature on a prepared sheet of photographic paper, which is carried on horizontally behind the instrument by means of clockwork. In the same manner the barometric indications are traced; and an entirely new feature is introduced by the photographic registration of the amount of rain and the time of fall.

The rain from the funnel is received in one leg of a siphon tube, and the weight of the water causes a column of mercury to ascend in the other; this column shuts out the light and records the amount received, in the same way as the mercury acts in the barometer and thermometer.

As the photographic registers are all reduced to the same size by the intervention of lenses, positive copies of any or all of the negative originals may be readily taken on one sheet of paper, and multiplied to any extent; we are thus enabled to see at once the connexion between the thermometric, barometric, and hygrometric curves, impressed by nature without the trouble of geometrical construction. The manner in which these combined curves are presented to the eye at a glance is most striking, and they indicate most important relations between various meteorological phenomena which have yet to be fully developed.

At Cambridge, Liverpool, and some other places few in number, very valuable and efficient observers have recorded and published registrations more extensive than can be expected from private observers, but far inferior to the elaborate system pursued at Greenwich. Private individuals in various localities record phenomena without publishing their results, or joining any Society which has the cultivation of meteorological science in view. I apprehend their number is not great, or they would be more generally known.

From the view of what is accomplished by extra-observatorial efforts, we are compelled to arrive at the conclusion that much remains to be done before we shall become acquainted with simultaneous movements in the air, or variations in its thermo-hygrometric state, even within so narrow a district as our own country. We shall now, as a consequence of the late movement, enlist on our side the officers of the Royal Navy and of the Mercantile Marine. The ships of Great Britain traverse the ocean in every direction, and at all periods of the year; they are commanded by men accustomed to watch natural phenomena, and the regularity of life at sea is favourable to the systematic registration of the barometer and thermometer: to render the records valuable, the instruments will henceforth be of a superior character to those usually found on ship-board, calculated not only to show differential but absolute values.

Those observers whose reports are published every three months at present, from want more of time than inclination, confine their observations within too narrow a range. In addition to the pressure, temperature, and hygrometric state of the air, it would be highly advantageous could we, for all localities, ascertain in addition the rapidity of evaporation, the range and intensity of solar radiation, and the state of electric tension; all which, in their varied combinations, go to make up that general result which we call *climate*, and which, unitedly, produce effects upon the natural world and the human frame, varying according to the preponderance of one or the other element. A knowledge of all these would lead us, most probably, to conclusions approaching the truth as to the adaptation of one particular series of crops to certain parts of the kingdom, and of the fitness of certain places for those who are suffering from peculiar diseases. We do not, moreover, at present distinguish the rainy hours in a day, but simply record the daily fall; and this leaves us deficient in one important element. Upon the whole, we may conclude that meteorological science is in a state of infancy; that it is, and must long continue to be, only a science of observation; that recorded phenomena are at present too few, and those taken over only a small portion of the earth's surface;—nay, the two-thirds of that surface occupied by the ocean, though exercising a most important influence on atmospheric changes, may, as regards correct observation, be considered as till lately a blank—too few are they and insignificant to enable us to draw conclusions or deductions which shall hold good over a large extent.

Meteorology is precisely in that position in which geology was found eighty years ago, or microscopic science at a still later period; and yet, since that time, how many facts then obscure have been elucidated in the structure of the earth! for how many sound principles has geology gained universal reception! How many secrets of nature has the microscope disclosed! how many wonderful processes of nature has it unveiled!

APPENDIX.

Table I.—Correspondence of the different Thermometrical scales.

TABLE FOR THE CENTIGRADE THERMOMETER.

Centigrade.	Reaumur's.	Fahrenheit's.	Centigrade.	Reaumur's.	Fahrenheit's.
100	80·	212·	73	58·4	163·4
99	79·2	210·2	72	57·6	161·6
98	78·4	208·4	71	56·8	159·8
97	77·6	206·6	70	56·	158·
96	76·8	204·8	69	55·2	156·2
95	76·	203·	68	54·4	154·4
94	75·2	201·2	67	53·6	152·6
93	74·4	199·4	66	52·8	150·8
92	73·6	197·6	65	52·	149·
91	72·8	195·8	64	51·2	147·2
90	72·	194·	63	50·4	145·4
89	71·2	192·2	62	49·6	143·6
88	70·4	190·4	61	48·8	141·8
87	69·6	188·6	60	48·	140·
86	68·8	186·8	59	47·2	138·2
85	68·	185·	58	46·4	136·4
84	67·2	183·2	57	45·6	134·6
83	66·4	181·4	56	44·8	132·8
82	65·6	179·6	55	44·	131·
81	64·8	177·8	54	43·2	129·2
80	64·	176·	53	42·4	127·4
79	63·2	174·2	52	41·6	125·6
78	62·4	172·4	51	40·8	123·8
77	61·6	170·6	50	40·	122·
76	60·8	168·8	49	39·2	120·2
75	60·	167·	48	38·4	118·4
74	59·2	165·2	47	37·6	116·6

TABLE for the Centigrade Thermometer (*continued*).

Centigrade.	Reaumur's.	Fahrenheit's.	Centigrade.	Reaumur's.	Fahrenheit's.
46	36·8	114·8	7	5·6	44·6
45	36·	113·	6	4·8	42·8
44	35·2	111·2	5	4·	41·
43	34·4	109·4	4	3·2	39·2
42	33·6	107·6	3	2·4	37·4
41	32·8	105·8	2	1·6	35·6
40	32·	104·	1	0·8	33·8
39	31·2	102·2	0	0·	32·
38	30·4	100·4	− 1	− 0·8	30·2
37	29·6	98·6	− 2	− 1·6	28·4
36	28·8	96·8	− 3	− 2·4	26·6
35	28·	95·	− 4	− 3·2	24·8
34	27·2	93·2	− 5	− 4·	23·
33	26·4	91·4	− 6	− 4·8	21·2
32	25·6	89·6	− 7	− 5·6	19·4
31	24·8	87·8	− 8	− 6·4	17·6
30	24·	86·	− 9	− 7·2	15·8
29	23·2	84·2	−10	− 8·	14·
28	22·4	82·4	−11	− 8·8	12·2
27	21·6	80·6	−12	− 9·6	10·4
26	20·8	78·8	−13	−10·4	8·6
25	20·	77·	−14	−11·2	6·8
24	19·2	75·2	−15	−12·	5·
23	18·4	73·4	−16	−12·8	3·2
22	17·6	71·6	−17	−13·6	1·4
21	16·8	69·8	−18	−14·4	− 0·4
20	16·	68·	−19	−15·2	− 2·2
19	15·2	66·2	−20	−16·	− 4·
18	14·4	64·4	−21	−16·8	− 5·8
17	13·6	62·6	−22	−17·6	− 7·6
16	12·8	60·8	−23	−18·4	− 9·4
15	12·	59·	−24	−19·2	−11·2
14	11·2	57·2	−25	−20·	−13·
13	10·4	55·4	−26	−20·8	−14·8
12	9·6	53·6	−27	−21·6	−16·6
11	8·8	51·8	−28	−22·4	−18·4
10	8·	50·	−29	−23·2	−20·2
9	7·2	48·2	−30	−24·	−22·
8	6·4	46·4	−31	−24·8	−23·8

TABLE for the Centigrade Thermometer (*continued*).

Centigrade.	Reaumur's.	Fahrenheit's.	Centigrade.	Reaumur's.	Fahrenheit's.
−32	−25·6	−25·6	−37	−29·6	−34·6
−33	−26·4	−27·4	−38	−30·4	−36·4
−34	−27·2	−29·2	−39	−31·2	−38·2
−35	−28·	−31·	−40	−32·	−40·
−36	−28·8	−32·8			

TABLE FOR REAUMUR'S THERMOMETER.

Reaumur's.	Centigrade.	Fahrenheit's.	Reaumur's.	Centigrade.	Fahrenheit's.
80	100·	212·	54	67·5	153·5
79	98·75	209·75	53	66·25	151·25
78	97·5	207·5	52	65·	149·
77	96·25	205·25	51	63·75	146·75
76	95·	203·	50	62·5	144·5
75	93·75	200·75	49	61·25	142·25
74	92·5	198·5	48	60·	140·
73	91·25	196·25	47	58·75	137·75
72	90·	194·	46	57·5	135·5
71	88·75	191·75	45	56·25	133·25
70	87·5	189·5	44	55·	131·
69	86·25	187·25	43	53·75	128·75
68	85·	185·	42	52·5	126·5
67	83·75	182·75	41	51·25	124·25
66	82·5	180·5	40	50·	122·
65	81·25	178·25	39	48·75	119·75
64	80·	176·	38	47·5	117·5
63	78·75	173·75	37	46·25	115·25
62	77·5	171·5	36	45·	113·
61	76·25	169·25	35	43·75	110·75
60	75·	167·	34	42·5	108·5
59	73·75	164·75	33	41·25	106·25
58	72·5	162·5	32	40·	104·
57	71·25	160·25	31	38·75	101·75
56	70·	158·	30	37·5	99·5
55	68·75	155·75	29	36·25	97·25

TABLE for Reaumur's Thermometer (*continued*).

Reaumur's.	Centigrade.	Fahrenheit's.	Reaumur's.	Centigrade.	Fahrenheit's.
28	35·	95·	− 3	− 3·75	25·25
27	33·75	92·75	− 4	− 5·	23·
26	32·5	90·5	− 5	− 6·25	20·75
25	31·25	88·25	− 6	− 7·5	18·5
24	30·	86·	− 7	− 8·75	16·25
23	28·75	83·75	− 8	−10·	14·
22	27·5	81·5	− 9	−11·25	11·75
21	26·25	79·25	−10	−12·5	9·5
20	25·	77·	−11	−13·75	7·25
19	23·75	74·75	−12	−15·	5·
18	22·5	72·5	−13	−16·25	2·75
17	21·25	70·25	−14	−17·5	0·5
16	20·	68·	−15	−18·75	− 1·75
15	18·75	65·75	−16	−20·	− 4·
14	17·5	63·5	−17	−21·25	− 6·25
13	16·25	61·25	−18	−22·5	− 8·5
12	15·	59·	−19	−23·75	−10·75
11	13·75	56·75	−20	−25·	−13·
10	12·5	54·5	−21	−26·25	−15·25
9	11·25	52·25	−22	−27·5	−17·5
8	10·	50·	−23	−28·75	−19·75
7	8·75	47·75	−24	−30·	−22·
6	7·5	45·5	−25	−31·25	−24·25
5	6·25	43·25	−26	−32·5	−26·5
4	5·	41·	−27	−33·75	−28·75
3	3·75	38·75	−28	−35·	−31·
2	2·5	36·5	−29	−36·25	−33·25
1	1·25	34·25	−30	−37·5	−35·5
0	0·	32·	−31	−38·75	−37·75
−1	−1·25	29·75	−32	−40·	−40·
−2	−2·5	27·5	−33	−41·25	−42·25

TABLE FOR FAHRENHEIT'S THERMOMETER*.

Fahrenheit's.	Reaumur's.	Centigrade.	Fahrenheit's.	Reaumur's.	Centigrade.
212	80·00	100·00	176	64·00	80·00
211	79·55	99·44	175	63·55	79·44
210	79·11	98·88	174	63·11	78·88
209	78·66	98·33	173	62·66	78·33
208	78·22	97·77	172	62·22	77·77
207	77·77	97·22	171	61·77	77·22
206	77·33	96·66	170	61·33	76·66
205	76·88	96·11	169	60·88	76·11
204	76·44	95·55	168	60·44	75·55
203	76·00	95·00	167	60·00	75·00
202	75·55	94·44	166	59·55	74·44
201	75·11	93·88	165	59·11	73·88
200	74·66	93·33	164	58·66	73·33
199	74·22	92·77	163	58·22	72·77
198	73·77	92·22	162	57·77	72·22
197	73·33	91·66	161	57·33	71·66
196	72·88	91·11	160	56·88	71·11
195	72·44	90·55	159	56·44	70·55
194	72·00	90·00	158	56·00	70·00
193	71·55	89·44	157	55·55	69·44
192	71·11	88·88	156	55·11	68·88
191	70·66	88·33	155	54·66	68·33
190	70·22	87·77	154	54·22	67·77
189	69·77	87·22	153	53·77	67·22
188	69·33	86·66	152	53·33	66·66
187	68·88	86·11	151	52·88	66·11
186	68·44	85·55	150	52·44	65·55
185	68·00	85·00	149	52·00	65·00
184	67·55	84·44	148	51·55	64·44
183	67·11	83·88	147	51·11	63·88
182	66·66	83·33	146	50·66	63·33
181	66·22	82·77	145	50·22	62·77
180	65·77	82·22	144	49·77	62·22
179	65·33	81·66	143	49·33	61·66
178	64·88	81·11	142	48·88	61·11
177	64·44	80·55	141	48·44	60·55

* All the decimals in this Table are circulating decimals.

TABLE for Fahrenheit's Thermometer (*continued*).

Fahrenheit's.	Reaumur's.	Centigrade.	Fahrenheit's.	Reaumur's.	Centigrade.
140	48·00	60·00	101	30·66	38·33
139	47·55	59·44	100	30·22	37·77
138	47·11	58·88	99	29·77	37·22
137	46·66	58·33	98	29·33	36·66
136	46·22	57·77	97	28·88	36·11
135	45·77	57·22	96	28·44	35·55
134	45·33	56·66	95	28·00	35·00
133	44·88	56·11	94	27·55	34·44
132	44·44	55·55	93	27·11	33·88
131	44·00	55·00	92	26·66	33·33
130	43·55	54·44	91	26·22	32·77
129	43·11	53·88	90	25·77	32·22
128	42·66	53·33	89	25·33	31·66
127	42·22	52·77	88	24·88	31·11
126	41·77	52·22	87	24·44	30·55
125	41·33	51·66	86	24·00	30·00
124	40·88	51·11	85	23·55	29·44
123	40·44	50·55	84	23·11	28·88
122	40·00	50·00	83	22·66	28·33
121	39·55	49·44	82	22·22	27·77
120	39·11	48·88	81	21·77	27·22
119	38·66	48·33	80	21·33	26·66
118	38·22	47·77	79	20·88	26·11
117	37·77	47·22	78	20·44	25·55
116	37·33	46·66	77	20·00	25·00
115	36·88	46·11	76	19·55	24·44
114	36·44	45·55	75	19·11	23·88
113	36·00	45·00	74	18·66	23·33
112	35·55	44·44	73	18·22	22·77
111	35·11	43·88	72	17·77	22·22
110	34·66	43·33	71	17·33	21·66
109	34·22	42·77	70	16·88	21·11
108	33·77	42·22	69	16·44	20·55
107	33·33	41·66	68	16·00	20·00
106	32·88	41·11	67	15·55	19·44
105	32·44	40·55	66	15·11	18·88
104	32·00	40·00	65	14·66	18·33
103	31·55	39·44	64	14·22	17·77
102	31·11	38·88	63	13·77	17·22

TABLE for Fahrenheit's Thermometer (*continued*).					
Fahrenheit's.	Reaumur's.	Centigrade.	Fahrenheit's.	Reaumur's.	Centigrade.
62	13·33	16·66	23	−4·00	−5·00
61	12·88	16·11	22	−4·44	−5·55
60	12·44	15·55	21	−4·88	−6·11
59	12·00	15·00	20	−5·33	−6·66
58	11·55	14·44	19	−5·77	−7·22
57	11·11	13·88	18	−6·22	−7·77
56	10·66	13·33	17	−6·66	−8·33
55	10·22	12·77	16	−7·11	−8·88
54	9·77	12·22	15	−7·55	−9·44
53	9·33	11·66	14	−8·00	−10·00
52	8·88	11·11	13	−8·44	−10·55
51	8·44	10·55	12	−8·88	−11·11
50	8·00	10·00	11	−9·33	−11·66
49	7·55	9·44	10	−9·77	−12·22
48	7·11	8·88	9	−10·22	−12·77
47	6·66	8·33	8	−10·66	−13·33
46	6·22	7·77	7	−11·11	−13·88
45	5·77	7·22	6	−11·55	−14·44
44	5·33	6·66	5	−12·00	−15·00
43	4·88	6·11	4	−12·44	−15·55
42	4·44	5·55	3	−12·88	−16·11
41	4·00	5·00	2	−13·33	−16·66
40	3·55	4·44	1	−13·77	−17·22
39	3·11	3·88	0	−14·22	−17·77
38	2·66	3·33	−1	−14·66	−18·33
37	2·22	2·77	−2	−15·11	−18·88
36	1·77	2·22	−3	−15·55	−19·44
35	1·33	1·66	−4	−16·00	−20·00
34	0·88	1·11	−5	−16·44	−20·55
33	0·44	0·55	−6	−16·88	−21·11
32	0·	0·	−7	−17·33	−21·66
31	−0·44	−0·55	−8	−17·77	−22·22
30	−0·88	−1·11	−9	−18·22	−22·77
29	−1·33	−1·66	−10	−18·66	−23·33
28	−1·77	−2·22	−11	−19·11	−23·88
27	−2·22	−2·77	−12	−19·55	−24·44
26	−2·66	−3·33	−13	−20·00	−25·00
25	−3·11	−3·88	−14	−20·44	−25·55
24	−3·55	−4·44	−15	−20·88	−26·11

TABLE for Fahrenheit's Thermometer (*continued*).

Fahrenheit's.	Reaumur's.	Centigrade.	Fahrenheit's.	Reaumur's.	Centigrade.
−16	−21·33	−26·66	−29	−27·11	−33·88
−17	−21·77	−27·22	−30	−27·55	−34·44
−18	−22·22	−27·77	−31	−28·00	−35·00
−19	−22·66	−28·33	−32	−28·44	−35·55
−20	−23·11	−28·88	−33	−28·88	−36·11
−21	−23·55	−29·44	−34	−29·33	−36·66
−22	−24·00	−30·00	−35	−29·77	−37·22
−23	−24·44	−30·55	−36	−30·22	−37·77
−24	−24·88	−31·11	−37	−30·66	−38·33
−25	−25·33	−31·66	−38	−31·11	−38·88
−26	−25·77	−32·22	−39	−31·55	−39·44
−27	−26·22	−32·77	−40	−32·00	−40·00
−28	−26·66	−33·33			

TABLE II.—Tension, or Elastic Force, of Aqueous Vapour in inches of mercury, for every degree of temperature from 0° to 95°.

Temp.	Tension.	Temp.	Tension.	Temp.	Tension.	Temp.	Tension.
0°	·044	12°	·074	24°	·129	36°	·212
1	·046	13	·078	25	·135	37	·220
2	·048	14	·082	26	·141	38	·229
3	·050	15	·086	27	·147	39	·238
4	·052	16	·090	28	·153	40	·247
5	·054	17	·094	29	·160	41	·257
6	·057	18	·098	30	·167	42	·267
7	·060	19	·103	31	·174	43	·277
8	·062	20	·108	32	·181	44	·288
9	·065	21	·113	33	·188	45	·299
10	·068	22	·118	34	·196	46	·311
11	·071	23	·123	35	·204	47	·323

TABLE II. (*continued*).

Temp.	Tension.	Temp.	Tension.	Temp.	Tension.	Temp.	Tension.
48°	·335	60°	·518	72°	·785	84°	1·165
49	·348	61	·537	73	·812	85	1·203
50	·361	62	·556	74	·840	86	1·242
51	·374	63	·576	75	·868	87	1·282
52	·388	64	·596	76	·897	88	1·323
53	·403	65	·617	77	·927	89	1·366
54	·418	66	·639	78	·958	90	1·410
55	·433	67	·661	79	·990	91	1·455
56	·449	68	·684	80	1·023	92	1·501
57	·465	69	·708	81	1·057	93	1·548
58	·482	70	·733	82	1·092	94	1·596
59	·500	71	·759	83	1·128	95	1·646

TABLE III.—The weight (in grains Troy) of vapour in a cubic foot of saturated air; at all temperatures between 10° and 89°.

Temp.	Weight.	Temp.	Weight.	Temp.	Weight.	Temp.	Weight.
10°	·8	25°	1·6	40°	2·9	55°	4·9
11	·9	26	1·7	41	3·0	56	5·0
12	·9	27	1·7	42	3·1	57	5·2
13	1·0	28	1·8	43	3·2	58	5·4
14	1·0	29	1·9	44	3·3	59	5·6
15	1·1	30	2·0	45	3·4	60	5·8
16	1·1	31	2·1	46	3·6	61	6·0
17	1·1	32	2·1	47	3·7	62	6·2
18	1·2	33	2·2	48	3·8	63	6·4
19	1·3	34	2·3	49	4·0	64	6·6
20	1·3	35	2·4	50	4·1	65	6·8
21	1·4	36	2·5	51	4·2	66	7·0
22	1·4	37	2·6	52	4·4	67	7·3
23	1·5	38	2·7	53	4·5	68	7·5
24	1·5	39	2·8	54	4·7	69	7·8

TABLE III. (*continued*).

Temp.	Weight.	Temp.	Weight.	Temp.	Weight.	Temp.	Weight.
°		°		°		°	
70	8·0	75	9·4	80	11·0	85	12·8
71	8·3	76	9·7	81	11·3	86	13·2
72	8·5	77	10·0	82	11·7	87	13·6
73	8·8	78	10·3	83	12·0	88	14·0
74	9·1	79	10·6	84	12·4	89	14·4

TABLE IV.—Factors to be multiplied into the quantities in TABLE III., when the air-temperature and dew-point temperature differ by the number of degrees in the first column.

Diff.	Factor.	Diff.	Factor.	Diff.	Factor.	Diff.	Factor.
°		°		°		°	
1	·999	9	·982	17	·966	25	·951
2	·996	10	·980	18	·964	26	·949
3	·994	11	·978	19	·962	27	·947
4	·992	12	·976	20	·960	28	·945
5	·990	13	·974	21	·958	29	·943
6	·988	14	·972	22	·956	30	·942
7	·986	15	·970	23	·954	31	·939
8	·984	16	·968	24	·952	32	·937

TABLE V.—Showing the weight in grains Troy of a cubic foot of air saturated with moisture, under the pressure of 30 inches of mercury, at any temperature between 31° and 90°; and the excess of the weight of a cubic foot of dry air, under the same pressure, over that of a cubic foot of saturated air, also in grains, throughout the same range of temperature.

Temp.	Wt. of a cub. ft. of sat. air.	Excess.	Temp.	Wt. of a cub. ft. of sat. air.	Excess.
°			°		
31	566·8	1·2	61	531·7	3·5
32	565·6	1·2	62	530·6	3·6
33	564·4	1·3	63	529·4	3·8
34	563·2	1·3	64	528·3	3·8
35	562·0	1·4	65	527·1	4·0
36	560·8	1·4	66	526·0	4·1
37	559·6	1·5	67	524·9	4·2
38	558·4	1·6	68	523·7	4·4
39	557·2	1·7	69	522·6	4·5
40	556·0	1·7	70	521·4	4·7
41	554·9	1·7	71	520·3	4·8
42	553·7	1·8	72	519·1	5·0
43	552·5	1·9	73	518·0	5·1
44	551·4	1·9	74	516·8	5·4
45	550·2	2·0	75	515·7	5·5
46	549·0	2·1	76	514·6	5·6
47	547·9	2·1	77	513·4	5·8
48	546·7	2·2	78	512·3	6·0
49	545·5	2·3	79	511·1	6·2
50	544·4	2·4	80	510·0	6·3
51	543·2	2·5	81	508·8	6 6
52	542·1	2·5	82	507·7	6·7
53	540·9	2·7	83	506·5	7·0
54	539·8	2·7	84	505·3	7·2
55	538·6	2·8	85	504·2	7·4
56	537·5	2·9	86	503·0	7·7
57	536·3	3·1	87	501·9	7·8
58	535·2	3·1	88	500·7	8·1
59	534·0	3·3	89	499·6	8·3
60	532·8	3·4	90	498·4	8·6

TABLE VI.—Showing the degree of humidity of the atmosphere, deduced from the readings of the dry-bulb and wet-bulb thermometers, for the usual range occurring in England; complete saturation being 1.

D—W.	32°	33°	34°	35°	36°	37°	38°	39°	40°	41°
1°	·87	·89	·89	·90	·91	·91	·91	·92	·92	·92
2	·75	·78	·79	·80	·82	·83	·83	·84	·84	·84
3	·65	·69	·71	·72	·74	·75	·75	·77	·76	·77
4	·57	·61	·63	·65	·66	·68	·68	·70	·69	·70

	42°	43°	44°	45°	46°	47°	48°	49°	50°	51°
1	·92	·92	·92	·92	·93	·93	·93	·93	·93	·93
2	·85	·84	·84	·85	·86	·86	·86	·86	·86	·86
3	·78	·78	·77	·78	·79	·79	·79	·79	·80	·80
4	·72	·71	·71	·72	·73	·73	·73	·73	·74	·74
6	·60	·59	·59	·60	·61	·61	·62	·62	·63	·63
8	·49	·49	·49	·50	·51	·51	·52	·53	·53	·54

	52°	53°	54°	55°	56°	57°	58°	59°	60°	61°
1	·93	·93	·93	·93	·93	·93	·93	·94	·94	·94
2	·86	·86	·86	·87	·87	·87	·87	·88	·88	·88
3	·80	·80	·80	·81	·81	·81	·81	·82	·82	·82
4	·74	·74	·74	·75	·75	·75	·76	·76	·76	·77
6	·64	·64	·64	·65	·65	·65	·66	·66	·66	·67
8	·54	·55	·55	·56	·56	·57	·57	·57	·58	·58
10	·46	·47	·47	·48	·48	·49	·49	·49	·50	·50

	62°	63°	64°	65°	66°	67°	68°	69°	70°	71°
1	·94	·94	·94	·94	·94	·94	·94	·94	·94	·94
2	·88	·88	·88	·88	·88	·88	·88	·88	·88	·88
3	·82	·82	·82	·83	·83	·83	·83	·83	·83	·83
4	·77	·77	·77	·78	·78	·78	·78	·78	·78	·78
6	·67	·67	·67	·68	·68	·68	·68	·68	·69	·69
8	·58	·59	·59	·59	·60	·60	·60	·60	·61	·61
10	·50	·51	·51	·51	·52	·52	·52	·53	·53	·53
12	·44	·44	·45	·45	·45	·46	·46	·47	·47	·47

Table VI. (*continued*).

D—W.	72°	73°	74°	75°	76°	77°	78°	79°	80°	81°
°										
1	·94	·94	·94	·94	·94	·94	·94	·95	·95	·95
2	·89	·89	·89	·89	·89	·89	·89	·90	·90	·90
3	·84	·84	·84	·84	·84	·84	·84	·85	·85	·85
4	·79	·79	·79	·79	·79	·79	·79	·80	·80	·80
6	·69	·70	·70	·70	·71	·71	·71	·71	·71	·72
8	·61	·62	·62	·62	·63	·63	·63	·63	·63	·64
10	·54	·54	·55	·55	·55	·56	·56	·56	·56	·56
12	·48	·48	·48	·49	·49	·50	·50	·50	·50	·50

Explanation.—Enter the column headed D—W. with the difference between the readings of the dry-bulb and wet-bulb thermometers; ranging with it under the reading of the dry-bulb thermometer, found in one of the horizontal columns, will be the degree of humidity required.

Table VII.—Corrections to be subtracted from the readings of Barometers, *with brass scales* extending from the cistern to the top of the mercurial column, to reduce the observations to 32° Fahrenheit.

Temp.	Inches.						
	28	28·5	29	29·5	30	30·5	31
°							
29	·001	·001	·001	·001	·001	·001	·001
30	·004	·004	·004	·004	·004	·004	·004
31	·006	·006	·007	·007	·007	·007	·007
32	·009	·009	·009	·009	·009	·010	·010
33	·011	·012	·012	·012	·012	·012	·012

Table VII. (*continued*).							
Temp.	Inches.						
	28	28·5	29	29·5	30	30·5	31
° 34	·014	·014	·014	·015	·015	·015	·015
35	·016	·017	·017	·017	·018	·018	·018
36	·019	·019	·020	·020	·020	·021	·021
37	·021	·022	·022	·022	·023	·023	·024
38	·024	·024	·025	·025	·026	·026	·026
39	·026	·027	·027	·028	·028	·029	·029
40	·029	·029	·030	·030	·031	·031	·032
41	·031	·032	·033	·033	·034	·034	·035
42	·034	·034	·035	·036	·036	·037	·037
43	·036	·037	·038	·038	·039	·040	·040
44	·039	·040	·040	·041	·042	·042	·043
45	·041	·042	·043	·044	·044	·045	·046
46	·044	·045	·045	·046	·047	·048	·049
47	·046	·047	·048	·049	·050	·051	·051
48	·049	·050	·051	·052	·052	·053	·054
49	·051	·052	·053	·054	·055	·056	·057
50	·054	·055	·056	·057	·058	·059	·060
51	·056	·057	·058	·059	·060	·061	·062
52	·059	·060	·061	·062	·063	·064	·065
53	·061	·063	·064	·065	·066	·067	·068
54	·064	·065	·066	·067	·068	·070	·071
55	·066	·068	·069	·070	·071	·072	·073
56	·069	·070	·071	·073	·074	·075	·076
57	·071	·073	·074	·075	·076	·078	·079
58	·074	·075	·077	·078	·079	·081	·082
59	·076	·078	·079	·080	·082	·083	·085
60	·079	·080	·082	·083	·085	·086	·087
61	·081	·083	·084	·086	·087	·089	·090
62	·084	·085	·087	·088	·090	·091	·093
63	·086	·088	·089	·091	·093	·094	·096
64	·089	·090	·092	·094	·095	·097	·098
65	·091	·093	·095	·096	·098	·100	·101
66	·094	·096	·097	·099	·101	·102	·104
67	·096	·098	·100	·102	·103	·105	·107
68	·099	·101	·102	·104	·106	·108	·109
69	·101	·103	·105	·107	·109	·110	·112
70	·104	·106	·108	·109	·111	·113	·115

TABLE VII. (*continued*).

Temp.	Inches.						
	28	28·5	29	29·5	30	30·5	31
° 71	·106	·108	·110	·112	·114	·116	·118
72	·109	·111	·113	·115	·117	·119	·120
73	·111	·113	·115	·117	·119	·121	·123
74	·114	·116	·118	·120	·122	·124	·126
75	·116	·118	·120	·122	·125	·127	·129
76	·119	·121	·123	·125	·127	·129	·131
77	·121	·123	·126	·128	·130	·132	·134
78	·124	·126	·128	·130	·133	·135	·137
79	·126	·128	·131	·133	·135	·137	·140
80	·129	·131	·133	·136	·138	·140	·143
81	·131	·134	·136	·138	·141	·143	·145
82	·134	·136	·138	·141	·143	·146	·148
83	·136	·139	·141	·143	·146	·148	·151
84	·139	·141	·144	·146	·149	·151	·154
85	·141	·144	·146	·149	·151	·154	·156
86	·144	·146	·149	·151	·154	·156	·159
87	·146	·149	·151	·154	·157	·159	·162
88	·149	·151	·154	·157	·159	·162	·165
89	·151	·154	·156	·159	·162	·165	·167
90	·153	·156	·159	·162	·164	·167	·170

PLATE I.

PLATE II.

Pl. II.

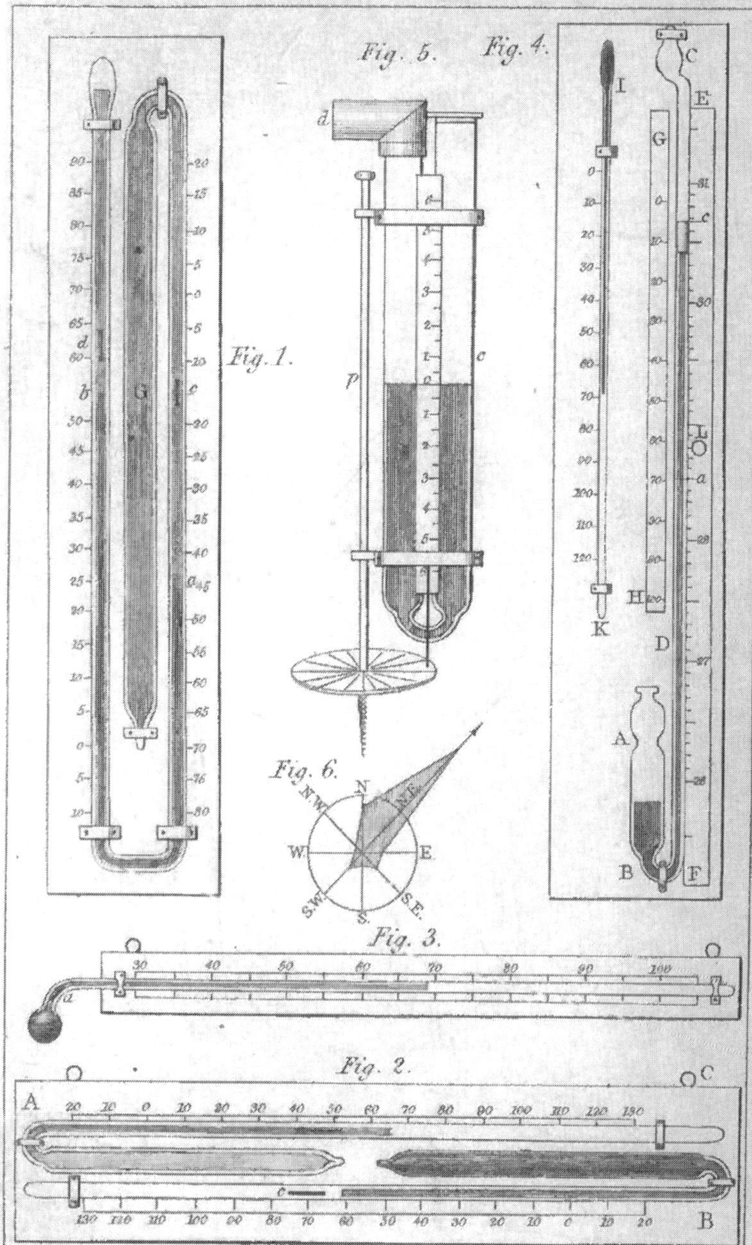

Fig. 1. Sixe's Register Thermometer. — 2. Rutherford's. — 3. Negretti's. 4. Adie's Sympiesometer. — 5. Lind's Wind-guage.

J. Basire. sc.

London: John Van Voorst, 1855.

PLATE III.

Curves of the monthly mean temperature at Greenwich and Southampton for the years 1848–1853, p. 78.

Pl. III.

Mean Monthly Temperatures at Greenwich and Southampton, for 6 years, compared.

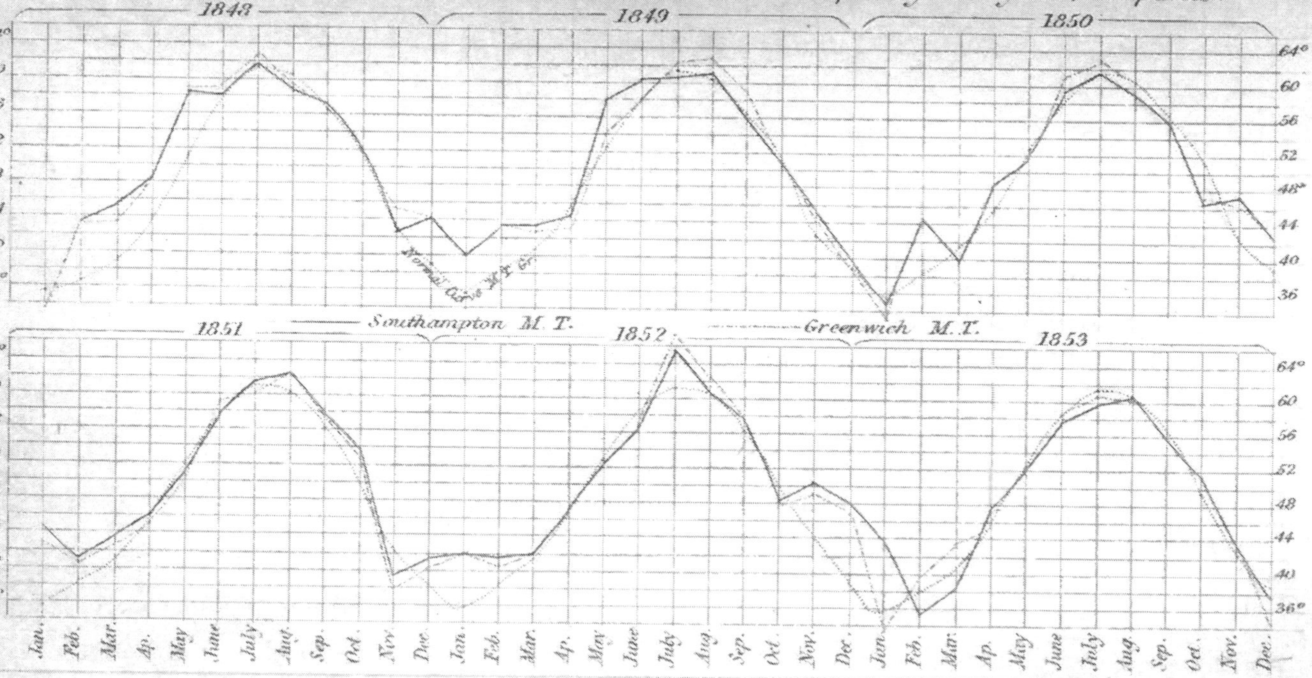

J. Basire sc.

PLATE IV.

Pl. IV.

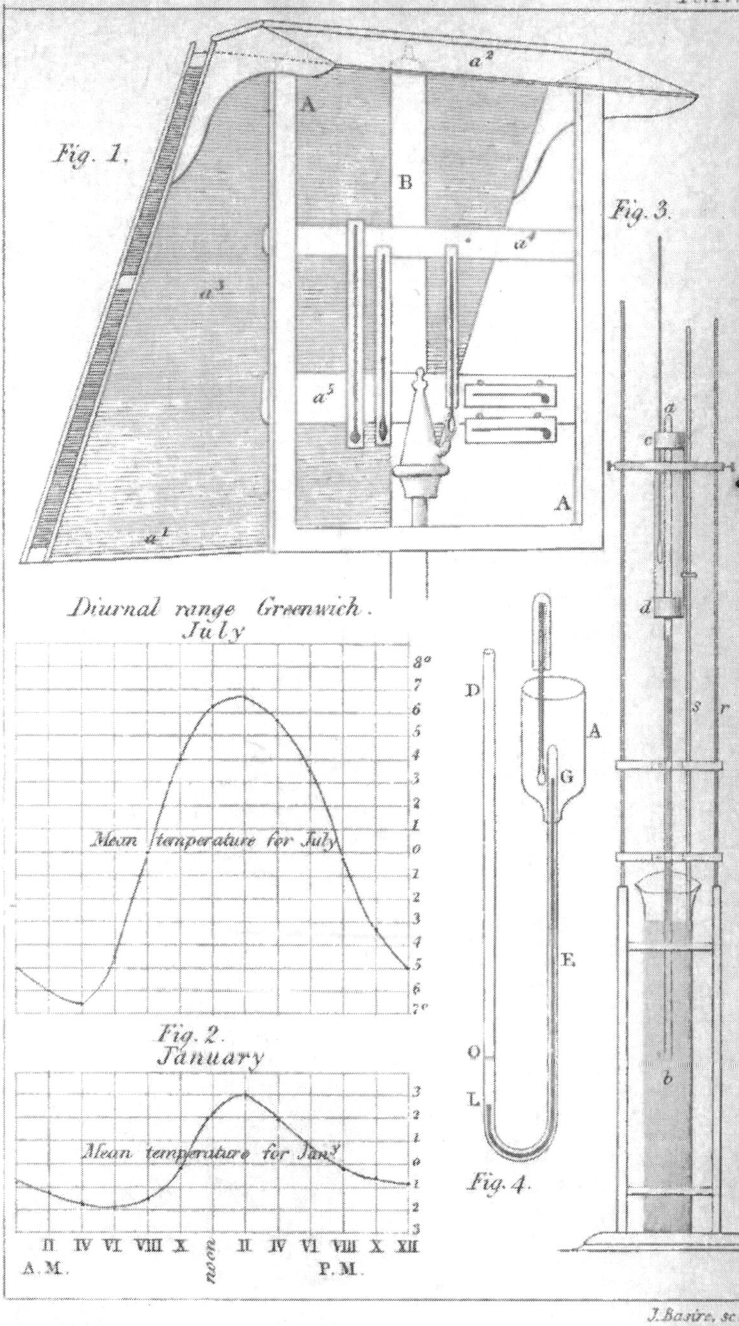

J. Basire, sc.

London: John Van Voorst, 1855.

PLATE V.

Curve of the mean annual temperature at Greenwich for the years 1771–1853, p. 84.

Curve of Mean Annual Temperature of the Air at the Royal Observatory at Greenwich, for every year from 1771 to 1853.

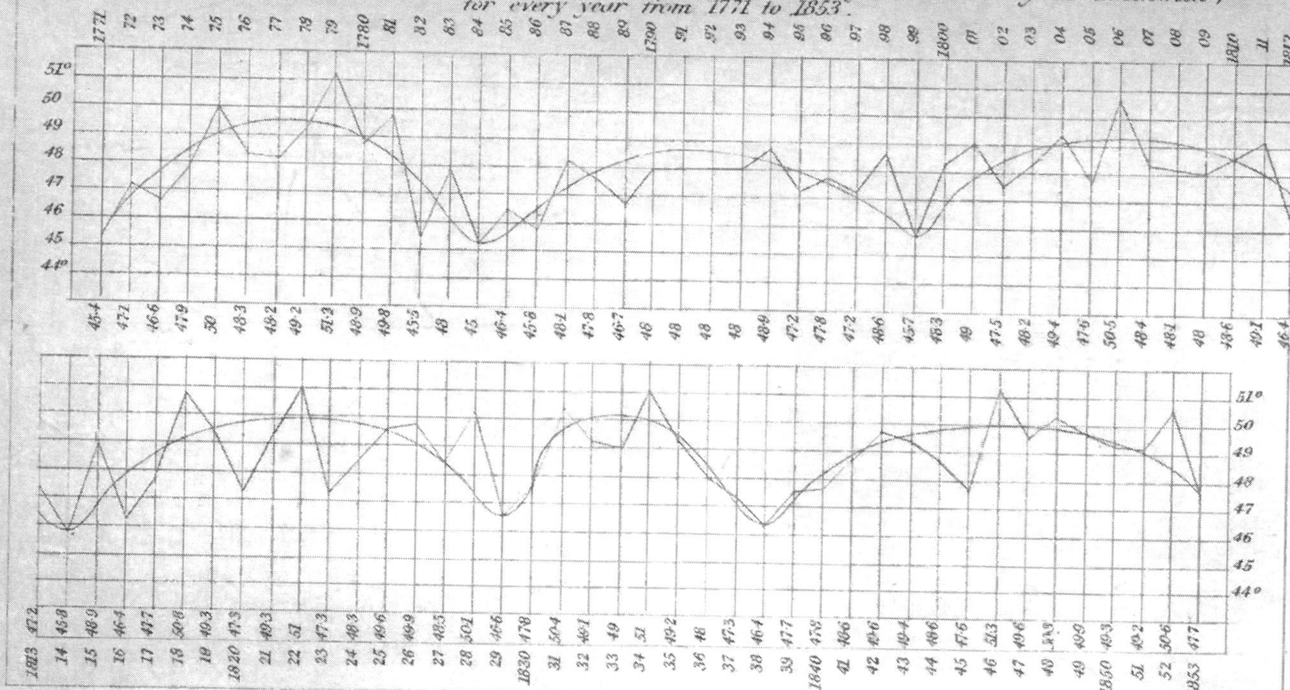

J. Basire, sc.

PLATE VI.

PLATE VII.

Sectional view of the Dome of the Kew Observatory, with the electrical instruments, p. 231.

Pl. VII.

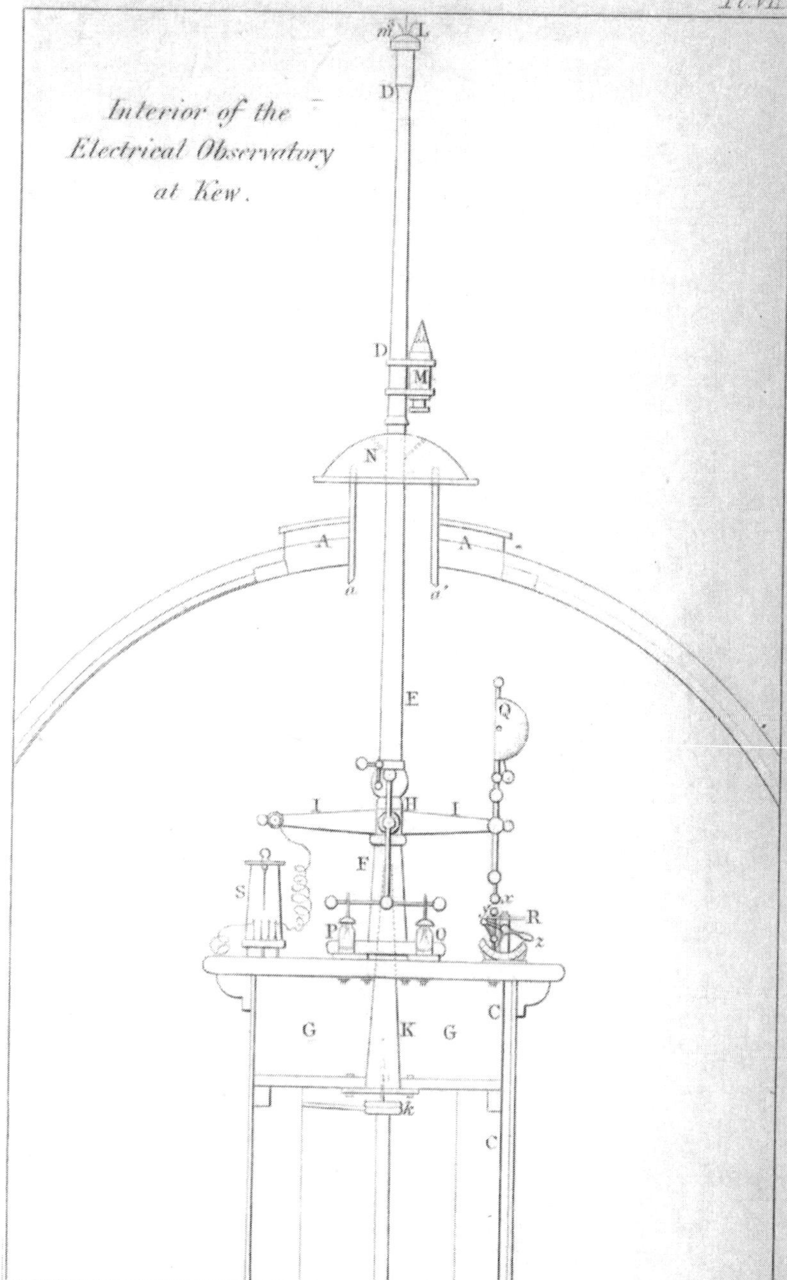

J. Basire. sc.

London: John Van Voorst, 1855.

PLATE VIII.

Fig. 1. Vertical cylinder with the photographic paper, p. 262.

Fig. 2. The paper unrolled, with the trace upon it for 12 hours, p. 263.

Figs. 3 and 4. Section and Plan of the Naphtha and Steam Box, p. 259.

Fig. 5. Registration of the movements of the Declination Magnet, pp. 264 and 267.

Fig. 6. Registration of the Barometer, pp. 266 and 268.

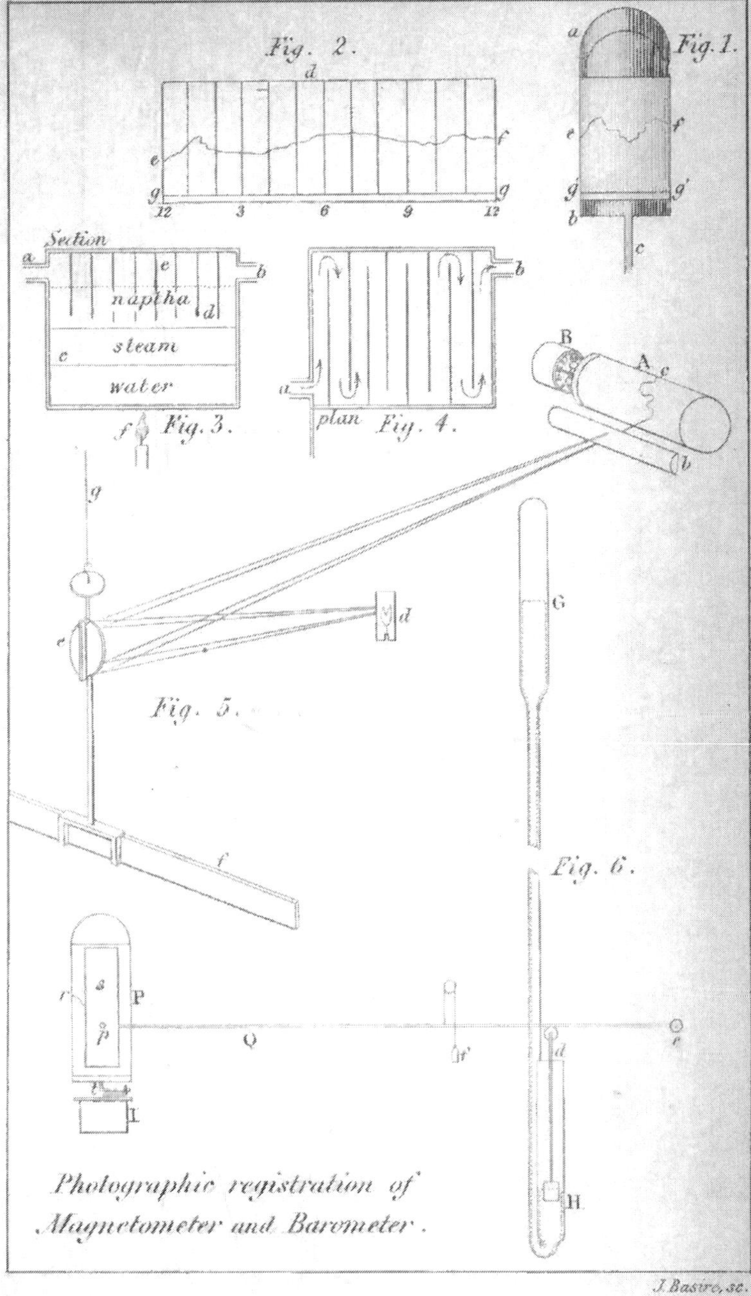

Photographic registration of Magnetometer and Barometer.

J. Basire, sc.

London: John Van Voorst, 1856.

PLATE IX.

Osler's Wind-gauge and Rain-gauge, with a specimen of its register, p. 272.

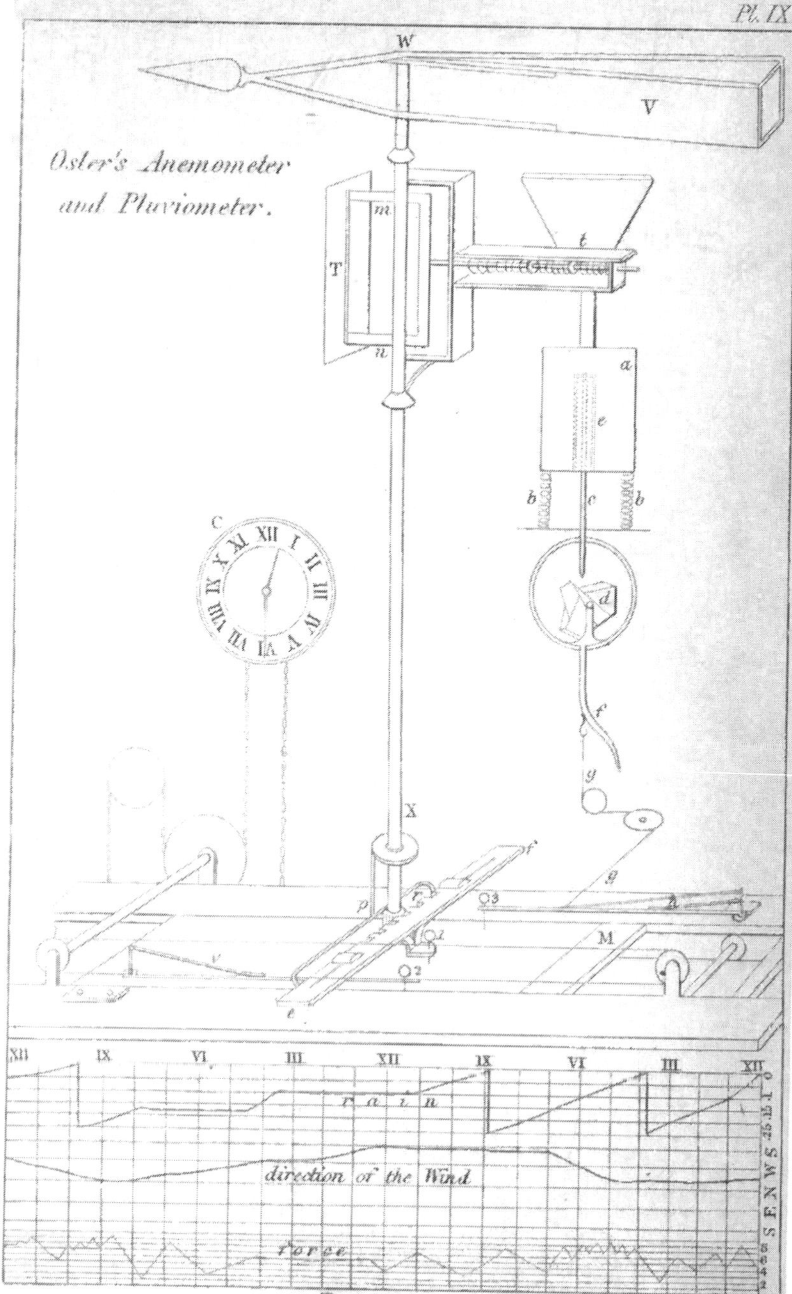

Register.

J. Basire, sc.

London: John Van Voorst, 1855.

PLATE X.

Whewell's Wind-gauge, p. 277.

Whewell's Anemometer.

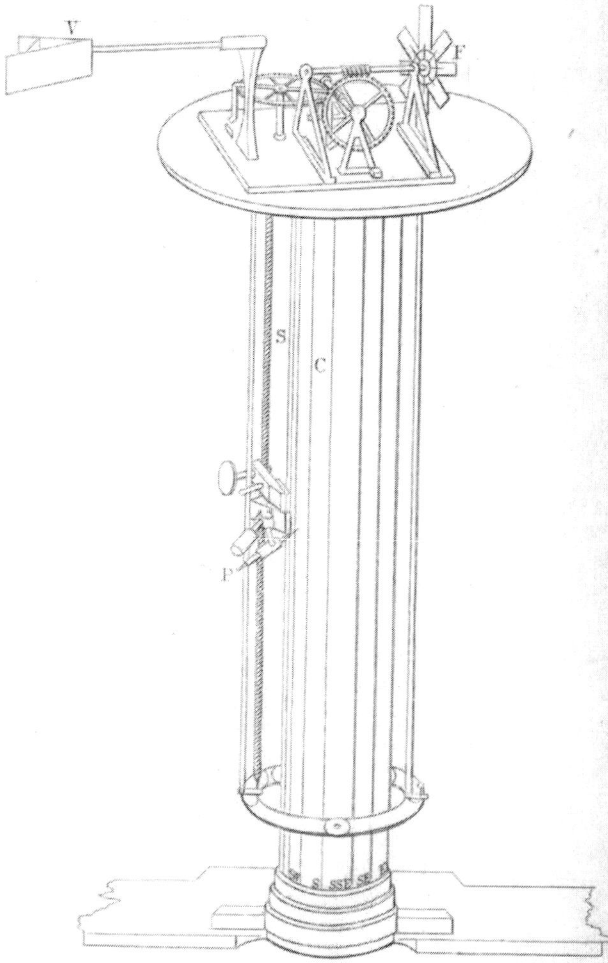

J. Basire sc.

London: John Van Voorst, 1855.

PLATE XI.

Fig. 1. Negretti and Zambra's Minimum Thermometer, p. 56.

Fig. 2. Robinson's Anemometer, p. 116.

Fig. 3. Hicks's Register Thermometer, p. 58.

Pl. XI.

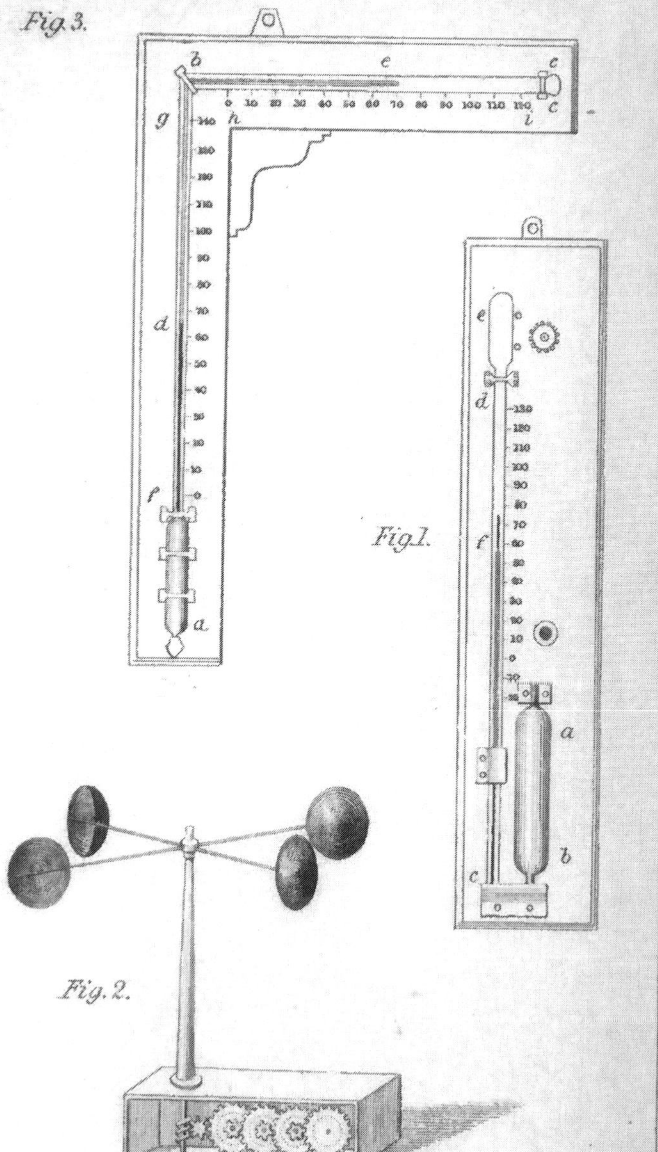

Fig. 1. Negretti & Zambra's Minimum Thermometer.—Fig. 2. Robinson's Wind guage.—Fig. 3. Hicks's Maximum & Minimum Thermometer.

J. Basire sc.

London: John Van Voorst, 1860.